AUERBACH ON
DATA COLLECTION SYSTEMS

AUERBACH ON
DATA COLLECTION SYSTEMS

AUERBACH®
publishers

princeton
philadelphia
new york
london

Library of Congress Catalog Card Number: 73-171050
International Standard Book Number: 0-87769-106-1

First Printing

Printed in the United States of America

E27155

CONTENTS

PREFACE

This volume is one of a series of books covering significant developments in the information science industry. *AUERBACH ON Data Collection Systems* covers both retail point-of-sale and industrial applications. Data collection is a simple concept; basically it is a method of transporting information from several locations to a central one. The concept becomes more restricted when used by computer-oriented people—"transporting" is replaced by "transmitting." Data is sent to a central location in an electronic, rather than physical, form. When used in this restricted context, a data collection system is defined as an *integrated network* of remotely located *specialized source data entry devices* with provisions for assembling *small units* of data at one central location. "Integrated network" means that the system components must be compatible and be able to communicate among themselves. "Source data entry" classifies those devices designed to receive data at the time and place of the data transaction and to convert the data into a machine-recognizable record. The remote devices should be either "specialized" for certain types of data transactions or simple to use, so that data entry is simple, straightforward, and subject to little chance for error. Because data is assumed to be entered at the source, it is entered as "small units" for convenience and efficiency, as opposed to being entered as batches of data. The most accepted use of the term "data collection" assumes that the collected information is transaction-oriented (based on individual events); thus the definition excludes such devices as batch terminals and data acquisition devices that obtain data exclusively from electronic sensors or monitors.

AUERBACH ON Data Collection is an expansion of material derived from *AUERBACH Data Handling* and *AUERBACH Data Communications Reports.* These publications are a major unit in the *AUERBACH Computer Technology Reports,* a looseleaf reference service recognized as the standard guide to EDP throughout the world. They are prepared and edited by the publisher's staff of EDP specialists.

Material in this volume was prepared by the staff of AUERBACH Publishers Inc. and has been updated prior to publication. However, due to the rapid changes occurring in the field, the currentness and completeness of its contents cannot be guaranteed. Information can be obtained from AUERBACH Publishers Inc., 121 North Broad Street, Philadelphia, Pa. 19107.

INTRODUCTION

At this stage in the development and use of computers, data collection systems are of great importance and value. Many businesses rely on computers but cannot use them efficiently because their methods of providing input to the computer are slow and sloppy. For those businesses with input problems centered on transaction-oriented data, data collection systems provide a solution.

Data collection is especially valuable in applications in which a computer-stored data base must be maintained and modified and updated as soon as possible after events occur that affect it. Following are some important applications of this type:

1. *Manufacturing*: job production status, tool control, attendance reporting, materials inspection, parts inventory.

2. *Retail Merchandising*: inventory status, customer account status and activity, salesperson activity.

3. *Wholesale Distribution*: merchandise receiving, order entry, stock status.

In each of these applications and in others, it might formerly have made little difference whether getting source data to the computer took minutes, hours, or days. But now it is becoming more and more the case that the time lag is indeed important. For example, a manufacturing executive inquiring of his data base or central information file about the status of certain jobs in production would prefer the data to be current *today*, not just as of last week; the same is true of the retail merchandiser inquiring about inventory status or the wholesaler about order status. For these people, data collection is a necessity. An advanced data collection system might even provide immediate access to an electronically stored data base, a particularly valuable feature when decisions or actions must be based on up-to-the-minute data.

One particular application, retail merchandising, is undergoing radical changes due to the many recent releases of data collection systems especially designed for the retailer. It is true that computers have automated much of the accounting and billing operations of stores, but computer technology has not been applied until recently, to any of the most basic operations, from the creation of data on the sales floor to and including the reporting of sales data in a computer-readable form. (Computer input has been the bottleneck.) Consequently, the computer has been isolated, both in time and place, from what is really going on in a store.

In only the past four years, however, computer manufacturers have been concentrating on the retailers' needs, and some have recently developed data collection systems and devices that enable practically all retail operations to be aided by an integrated, computer-oriented system. So revolutionary is the equipment that it is replacing the conventional electro-mechanical cash register, which has become almost indigenous to the retail store.

The new retail equipment now makes it possible to collect and maintain a complete, current data base, and thus such hitherto impossible or impractical tasks as daily reporting of high-activity merchandise, inventory status, salesperson productivity, customer account balances, and financial status are now possible. Retailers can develop important statistics as events happen, not weeks afterward, and buy and market intelligently.

Other benefits derived from retail data collection systems are equally valuable: The terminals that replace the conventional cash register aid the salesperson's task in many ways, including automatically reading price and inventory data from a merchandise tag, calculating taxes, and printing sales slips; also, the retail systems can handle all phases of credit authorization automatically, drastically cutting fraud losses and reducing customer waiting time.

The wait-and-see attitude retailers adopted when the new retail systems were introduced is now over. Major retailers placed equipment orders in 1971 totaling over $200 million. Since this aspect of data collection—retail merchandising—represents most of the recent activity of data collection system manufacturers, this book treats data collection in two parts. The first part discusses retail data collection as a complete subject. It explains the need for it, how the first approaches were made, the introduction of retailing systems, the characteristics and capabilities of the new retailing systems, two offshoots of such systems, and what can be expected in the near future. The second part describes different implementations of data collection systems. It discusses parallels to the complete retail data collection systems in other areas and then discusses methods of implementing data collection by using special devices; specifically, techniques using new portable recording and data communications devices are explained. This approach should provide enough information to anyone having a potential use for data collection equipment. Those in the retailing fields will certainly benefit most directly, but the methods of implementing and designing retail data collection equipment have their parallels in all types of applications.

1. WHY RETAILERS NEED COMPUTERS

Retail merchandising is a field characterized by overwhelming amounts of data. In record-keeping alone, the retailer must maintain written accounts of inventory, financial status, sales, orders pending, vendor performance, and customer credit, among other items. Not only must he maintain these records, he must also be able to interpret and manipulate them so that he can analyze the status of his business, forecast trends, and make meaningful management decisions. In addition, the retailer must create a large number of original documents, from the preparation of sales receipts on the sales floor to the preparation of invoices, notices, and checks in the accounting office.

Looking at the retailer's problems, it is easy to comment that all he needs is a computer. Many retailers have acquired computers to process much of their work, but on closer inspection have found that the computers solve few of their problems. Although computers have the necessary power, retailers, like some users, have not fed and cared for them properly. Even at this stage, approaching the fourth generation of computer equipment, there are almost no truly integrated retailing computer installations.

Two important aspects of retailing—inventory and credit records—could be greatly aided by computer processing, but as yet they have been only partially automated.

INVENTORY

It is hard to conceive of retailers making intelligent buying and marketing decisions without knowing what inventory they have and how it is

1

changing. Yet most retailers work in this situation, not because they want to, but because they are unable to keep up-to-date records and analyses of their inventory, even with the aid of computer processing.

Maintaining data on inventory is difficult, since there is such a variety of merchandise classes, types, and styles. The great diversity in styles and colors of modern men's and women's clothing, for example, makes record-keeping of apparel items alone a monumental task.

Part of this book discusses the early applications of computer processing to this task. None of these approaches solved the problem, but they immediately began to attack the problem's core: data input. Actual processing of the data has not been a problem, but getting data from the warehouse, the stockroom, and especially the sales floor has created a complex situation. This book also describes most of the recent equipment that can automate data input.

The advantages of knowing one's inventory and its daily changes are tremendous. Armed with an accurate, up-to-date picture of his inventory, a retailer can fast-order merchandise that is selling quickly, reduce inventory, and stop orders on merchandise that is selling slowly; he can also begin to chart trends in customer needs, planning future buying and stocking accordingly. Such close control on inventory enables the retailer to cut his inventory considerably; with much less money invested in inventory, his profit margins increase. However, such advantages are possible only with a truly integrated approach to retail data processing.

CREDIT RECORDS

Customer credit is an extremely important aspect of retail merchandising. The most common application of computers to customer credit is the compilation and preparation of monthly invoices and the production of account histories and aged balances. In this application computers are used extensively and are an invaluable aid, saving a great deal of time and labor. However, control and maintenance of customer credit involves not only the preparation of such documents, but also the input and processing of charge transactions, changes in account status, and credit verification or authorization. Until recently, the use of computers in these areas has been limited. This book describes isolated computer applications in these areas and suggests systems that can handle all or nearly all the aspects of customer credit.

Since total system computerization moves the computer closer to the sales floor, the retailer gains important advantages. Computerization of credit verification enables the salesperson to check a customer's credit immediately so that transactions are faster, losses due to the use of stolen or forged cards are reduced, and the retailer can keep tabs on high-activity or overextended accounts and eliminate losses from fraudulent charges under the "floor limit." The inclusion of the processing of charge transactions in the computer systems permits daily and even more frequent automatic updating of customer accounts; consequently, he realizes smaller labor costs, more accurate bookkeeping, and earlier detection of the fraudulent use of accounts. Such advantages are forthcoming in retail merchandising, but it is still difficult to find a retailer with a truly complete computerized customer credit system.

2. APPROACHES TO THE INTEGRATION
OF RETAIL INFORMATION SYSTEMS

DATA INPUT

As has been noted, a great problem in retail computer systems has been getting data from the sales floor to the computer. In the past 20 years many approaches have been tried, some elementary and some a little more direct; practically all these methods are still in wide use.

Punched Cards

The most elementary method of entering sales data into a computer system is via punched cards. Typically, sales checks and/or merchandise tags are manually collected from the sales desks and transported to a keypunch area, where selected data on the documents is punched into tab cards at weekly intervals. Date, department number, total amount of sale, and customer charge number are commonly extracted from sales checks; department number, merchandise style code, vendor number, color, size, and price are commonly extracted from merchandise tags. The sales check data is input to customer charge compilations, and the merchandise tag data is input to inventory programs. If desired, inventory data can be derived from the sales checks by keypunching the appropriate data for each item listed on the check.

Several EDP service centers provide this service for stores that do not handle their own processing. These service centers often provide all phases of the data processing operation, from picking up the sales checks and/or merchandise tags to computing, preparing, and delivering invoices and management reports.

4

However, the punched card method of data input is the most indirect process available to the retailer. Sales checks that have been prepared by a salesperson, or cash register and merchandise tags prepared by the vendor or retailer, must be filed by the salesperson, manually collected, distributed, and transcribed onto punched cards. Thus the method requires much additional labor and is prone to inaccuracies, due to loss of original documents and errors in transcription. It is also slow. Much time is consumed in the manual handling of the original documents, the actual transcription, and the manual handling of the prepared tab cards.

Several of these steps could be eliminated by the use of portable key-punches, which are available in a wide variety of design and operating characteristics. The more sophisticated types are electrically driven, utilize a 10-key keyboard for numeric data input or a 12-key keyboard (arranged like a Touch-Tone telephone) for alphanumeric data input, and have a tabulating function for automatic field positioning.

Fig. 2-1. Varifab 404 portable key punch, with a Series 600 automatic input terminal, enables sales personnel to punch transaction data into tab cards, bypassing the keypunch rooms. Dials and a badge/card reader simplify data entry.

A model manufactured by Varifab (Fig. 2-1) incorporates a badge/card reading device, thus enabling a salesperson to record a charge transaction directly, for example, by inserting a department or sales person identification badge (if required) and the customer's charge card, and then keying in the total amount of the sale.

The direct recording of transactions via auxiliary devices (Fig. 2-2) reduces the processing time lag and the error rate, but increases the sales-

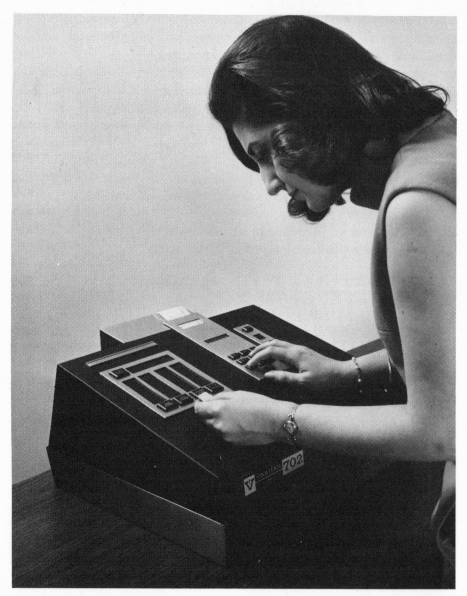

Fig. 2-2.   Varifab 702 data recorder enables sales personnel to record into punched paper tape the formated data for any of six types of transactions. Message lamps, word-length checks, and check-digit verification, each under transaction-dependent program control, guide and aid data entry.

person's responsibilities. It should also be noted that the direct recording of merchandise data, such as department code, vendor number, style, color, size, and price can become tedious for a salesperson. However, when confined to low- to medium-volume applications and to simple input procedures such as those for charge transactions, returns, or layaways, the use of portable key punches can be more advantageous than the use of dedicated key-punch operators.

Punched Merchandise Tags

The time-consuming and error-inducing procedure of manual merchandise tag transcription can be eliminated by the use of punched merchandise tags. These tags code inventory data such as stock-keeping unit (sku) number, vendor, color, style, size, and price in the form of punched holes. The same information is also printed on the tag. Their advantage over conventional tags is that they can be read by machines, avoiding the eye-to-hand step that bottlenecks many data processing systems.

Punched merchandise tags (Fig. 2-3) were introduced in the mid-1950s and are presently manufactured primarily by Kimball Systems and the

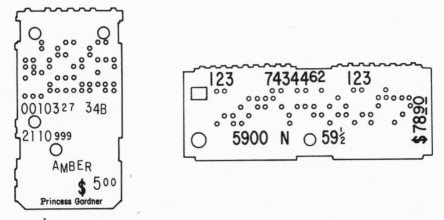

Fig. 2-3. Punched merchandise tags, primarily manufactured by Kimball and Dennison, are used widely to encode inventory data directly into a machine-readable medium, bypassing the manual keypunch step.

Dennison Manufacturing Company. They are used principally on garments, but can often be found on other dry goods and sometimes on different merchandise. Tag preparation is usually the responsibility of the retailer, who codes merchandise as part of the receiving procedure. However, some manufacturers supply their merchandise with prepunched tags affixed. Devices to encode the tags are usually electrically driven, and tag data is set by dials or input by punched cards or magnetic tape.

The initial portion of the data collection procedure for punched tags is much like that for conventional merchandise tags. As an item is sold, the tag is removed by the salesman and saved. At the end of the day, the tags are collected from the various sales areas and transported to a central location such as an EDP center. Processing, which is usually done weekly, can take many forms. Often a converter is used to transcribe the batch of tags onto punched cards on a one-to-one basis. The punched cards are then processed like key-punched cards for inventory. Other converters are available to transcribe the tags directly onto punched paper tape or magnetic tape. Many EDP service centers also provide processing services for punched tags.

The punched merchandise tag method of inventory data input has its problems too. Even though the manual transcription step is eliminated, errors and data processing delays occur. Tags must still be manually collected and transported, so it is inevitable that some tags are lost or damaged, making inventory reports inaccurate and incomplete. Furthermore, so much effort is involved in the manual handling and processing of tags that frequent reporting (more than once a week) is impractical and prohibitively expensive.

Problems resulting from tag-handling can be reduced with new tag-reading devices that have a data communications capability. A system called SPAN, announced by Kimball Systems in 1968, utilizes a tag-reading terminal to transmit tag data directly to a data processing center. The SPAN system reads batched tags and records the data on a magnetic-tape cassette that can be read immediately or subsequently, and the data is transmitted via telephone lines to a computer or a computer-compatible magnetic-tape recorder. Variable data can be added to tag batches by a keyboard on the tag reader.

Another system, manufactured by Binary Systems (Fig. 2-4), does much the same thing, but uses a specially printed tag instead of a punched tag.

Fig. 2-4. Binary Systems' bar-coded merchandise tag may be transcribed to another medium or the data read and sent to a remote location by using a terminal located near a cash register.

The tags are printed in a specialized bar-coded font with numeric symbols that are both man- and machine-readable. In one of its configurations, Binary Systems provides a reader/transmitter at each cash register location so that individual tags can be scanned and the data centrally recorded as soon as the tags are removed by the salesman.

Both the Kimball and Binary tag-reading/transmitting systems solve various problems inherent in merchandise tag systems, but their only product is computer input for inventory data; they furnish only merchandise-oriented data and do not provide sales-oriented data such as salesman number, customer account number, tax charged, and total amount of sale, all of which are necessary for a complete retail information system.

EDP-Oriented Cash Registers

A practical device for entering data into a retail information system is

Fig. 2-5. NCR Class 52 cash register records keyboard-entered and fixed data onto a punched paper tape or a journal tape printed in OCR font.

the common cash register. Unfortunately, cash registers are typically little more than adding machines with drawers. However, in the late 1950s, some manufacturers began to recognize the potential of cash registers as part of a data processing system and modified their cash registers to record sales transactions data directly onto a machine-readable medium. There are two classes of such devices: one punches a paper tape and one prints a journal tape in an OCR font (Fig. 2-5). Currently, the types that print OCR journal tapes are more popular. In contrast to merchandise tag methods, EDP-oriented cash registers can enter all data about a transaction into the EDP system.

Operation of these special cash registers is significantly different from that of conventional registers. The keyboard enters not only price data but also department code, salesperson identification, SKU or style number, and customer account number. All this data is punched into the paper tape or printed on the journal tape, with register number, store number, transaction number, and date automatically added by the register.

In a typical installation, printed journal or punched tapes are removed from each register at the end of each sales week and physically transported to an EDP center. The OCR tapes (Fig. 2-6) are usually directly processed by a computer and sometimes converted into a different input medium (punched cards, paper tape, or magnetic tape), while punched register tapes are typically converted into magnetic tape although they can be converted into some other input medium or processed directly if desired. The data can be used as input for a variety of data processing tasks, including customer billing, maintenance of credit files, inventory analysis, salesperson productivity, and many other forms of management reports.

Many EDP service centers provide processing services from punched or

```
0183840170041007N ⊣
018384678000006N ⊣
0183841230040005 ⊣
018374004800000S⊣ N ⊣
018374001000006N ⊣
0183740070000006N ⊣
```

Fig. 2-6.   The OCR-font journal tape constitutes cash register data directly recorded onto a machine-readable medium. Each line represents one keyboard-entered line plus some characters of fixed data. Tapes are normally processed by an EDP service center or by the store's EDP center if it contains an OCR journal tape reader.

Fig. 2-7. Farrington 4040 journal tape reader reads transaction data from an OCR-imprinted cash register journal tape and converts the data to magnetic tape or routes it to a computer.

printed register tapes, collecting tapes periodically and producing various types of reports and documents at specified intervals. One EDP service (Fig. 2-8), for example, prepares reports in the following categories:

1. Sales Audit: Summarizes transactions by employee, register, department, and store.

2. Merchandise Control: Reports inventory by unit, size, color, class, activity, and costs; reports monthly and yearly histories for comparison.

3. Accounts Receivable: Lists trial balances, delinquent accounts; computes customer historical analyses; prepares customer statements.

4. Vendor Analysis: Summarizes vendor performance, markdowns, and markups.

Many of the newer cash registers offer special capabilities in addition to punching a paper tape or printing an OCR font journal tape. Some of these features are the ability to read customer credit cards, enforce a predetermined sequence of operation, signal the operator if an entry error is made, and automatically prepare a daily register balance sheet. Other new

THE VILLAGE STORE

CUMULATIVE RETAIL INVENTORY MANAGEMENT REPORT   METHOD I

DEP NO.	CLASS NO.	UNITS SOLD	YTD SALES RETAIL	% OF SALES	YTD GR. PROFIT	% GROSS MARGIN	YTD COST PURCHASES	YTD RETAIL ADDITIONS	Y.T.D. MARKDOWNS	END INV AT COST	END INV AT RETAIL	% CUM MARKON
1	11	140	4,880.00	24.47	1,059.94	21.72	1,960.00	3,016.00	.00	6,251.94	7,986.00	21.72
1	12	62	800.00	4.01	204.10	25.51	1,000.00	1,450.00	64.00	2,604.10	3,776.00	31.03
1	13	58	1,600.00	8.02	464.64	29.04	1,920.00	2,706.00	.00	2,304.64	3,248.00	29.04
1	14	52	200.00	1.00	56.04	28.02	.00	.00	.00	136.04	189.00	28.02
DEPT		312	7,480.00	37.51	1,784.72	23.86	4,880.00	7,172.00	64.00	11,296.72	15,199.00	24.45
2	15	42	100.00	.50	31.94	31.94	.00	.00	.00	61.94	91.00	31.94
2	16	125	900.00	4.51	186.31	20.74	750.00	1,104.00	150.00	3,236.31	4,761.00	32.03
2	17	142	1,840.00	9.23	557.49	30.30	1,620.00	2,494.00	134.00	2,207.49	3,398.00	35.03
2	18	129	1,920.00	9.63	768.19	40.01	1,040.00	1,734.00	.00	1,608.19	2,681.00	40.01
2	19	138	200.00	1.00	78.20	39.10	180.00	296.00	.00	258.20	424.00	39.10
2	20	98	400.00	2.01	124.28	31.07	350.00	508.00	.00	474.28	688.00	31.07
DEPT		674	5,360.00	26.88	1,746.41	32.58	3,940.00	6,136.00	284.00	7,846.41	12,043.00	36.11
3	21	32	600.00	3.01	198.00	33.00	260.00	388.00	.00	548.00	818.00	33.00
3	22	70	300.00	1.50	95.91	31.97	290.00	426.00	.00	395.91	582.00	31.97
3	23	32	100.00	.50	34.07	34.07	150.00	228.00	.00	174.07	264.00	34.07
3	24	42	1,500.00	7.52	570.45	38.03	700.00	1,130.00	.00	670.45	1,082.00	38.03
3	25	185	3,700.00	18.56	1,332.74	36.02	1,820.00	2,844.00	.00	1,512.74	2,364.00	36.02
3	26	48	500.00	2.51	186.91	37.38	350.00	604.00	40.00	536.91	926.00	42.02
DEPT		409	6,700.00	33.60	2,418.08	36.09	3,570.00	5,620.00	40.00	3,838.08	6,036.00	36.44
INV.		12	400.00	2.01	119.80	29.95	320.00	456.00	.00	591.80	742.00	29.95
DEPT		12	400.00	2.01	119.80	29.95	320.00	456.00	.00	591.80	742.00	29.95
STORE		1407	19,940.00	100.00	6,069.01	30.44	12,710.00	19,384.00	388.00	23,501.01	34,020.00	31.72

Fig. 2-8.   Retail inventory report is prepared from cash register punched paper tapes by an NCR packaged program.

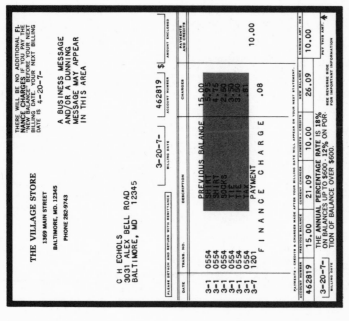

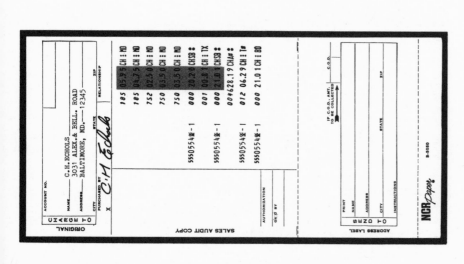

Fig. 2-9. Customer statement (right) is prepared from cash register punched paper tapes by an NCR packaged program. Corresponding areas on a register-printed sales check (left) and an audit journal tape (center) are shaded.

cash registers can record transaction data on a magnetic-tape cassette for later transcription to computer-compatible magnetic tape.

The EDP-oriented cash register is a powerful method of entering data in a retail information system, since a single salesclerk-operated device can produce machine-readable records of all required sales data. This device, like the others described above, has drawbacks, the most important of which is that a large amount of key entry is required to record a complete transaction, which often results in slower sales and in keying errors. On practically all registers of this type, the keyboard retains a conventional key arrangement, yet the cumbersome task of requiring two-line entries of typically 10 to 15 numbers each for every item sold would seem to call for a simpler keyboard or data entry arrangement. Some of the devices described in the following chapters do provide this convenience. In spite of the difficulty of keyboard entry and the delay resulting from manual handling of the register tapes, EDP-oriented cash registers provide a practical means of data entry with tolerable error rate, and are widely used by retailers today.

3. SYMBIOTIC RETAIL SYSTEMS

Cash registers designed to produce machine-readable output, as described in Chapter 2, were a first step in having the salesperson produce the medium for computer input of transaction data. This was an important step, since the salesperson's main function is to transact sales, and generation of all sales transaction data should also, ideally, be his responsibility. If his original creation of sales data also serves as data for computer input, an obvious savings is effected in the time and manpower required to tell a computer system what has been happening on the sales floor.

However, these cash registers were only an elementary approach and did little to solve the problem of designing an effective man/machine interface. To enter sales transaction data, the salesperson had to follow a complicated procedure and use a reference book to remember the steps for different types of transactions. The registers were modified only to enable a computer to read what was entered, retaining basically the same keyboard and operating mechanism; thus no symbiotic relationship was developed between man and computer.

In 1965 UNI-TOTE (a division of the American Totalisator Company) introduced a retail information system that could truly be called a point-of-sale (POS) system; that is, a system designed specifically for information system data entry by the salesperson as he handles a sales transaction. The system was accepted slowly, owing to the general reluctance of retailers to accept new methods, but it had still gained enough acceptance by early 1971 to be considered the most successful POS system. Major installations at that time included Gimbels in Pittsburgh, Joseph Magnins in San Francisco, and D. H. Holmes in New Orleans. The basic system had generally remained the same until mid-1971, when UNI-TOTE introduced its 300 system, which comprised electronic cash registers and programmable minicomputer controllers.

The original retail system of UNI-TOTE is basically configured with up to 128 terminals, or POS registers (Fig. 3-1), connected via a cable or communications line to a central control/receiving unit. Two aspects of the system indicate that its designers considered both the man/machine and the machine/computer system interface.

Fig. 3-1. The UNI-TOTE point-of-sale register broke cash register tradition by transferring transaction data to a centrally located magnetic-tape recorder, using illuminated messages to guide the operator, and performing credit verification.

First, the registers are designed to simplify data entry and to reduce the frequency of keying errors. The keyboard (Fig. 3-2) is a simplified version of a conventional cash register keyboard, with 11 columns of 9 numeric keys each. Data entry is guided by the selective illumination of an indicator panel above the category of data that should be entered into each key column, and the system indicates the set of control keys that are valid for a particular transaction stage.

Operating procedures are further simplified by automatic handling and

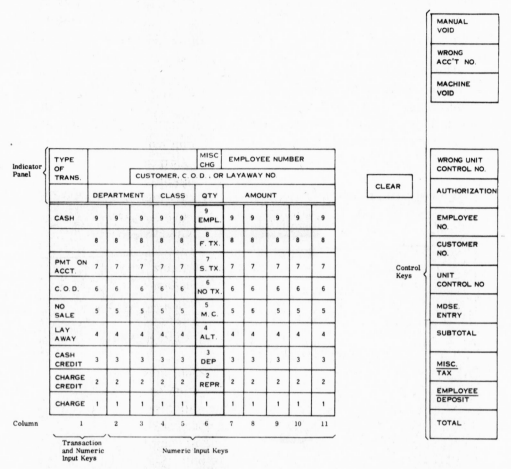

Fig. 3-2. The UNI-TOTE keyboard uses an indicator panel and illuminated control keys to help the salesperson follow the proper operating sequence.

printing of sales slips. Multipart fan-folded sales slips are automatically fed into the printer at the start of a transaction and automatically ejected at the end. All data entered by the operator (transaction type, employee number, customer charge number, merchandise department number, classification, quantity, and price; miscellaneous charge codes; and tax amount) is printed on the sales slip (Fig. 3-3), as is data automatically provided by the register (date, store number, register number, transaction number, subtotal, and total). The security of the audit tape of conventional cash registers is retained because one copy of each sales slip is stored and locked into the register.

Data entry errors are reduced by lamp illumination procedures designed to guide data entry and by a keyboard interlock system that en-

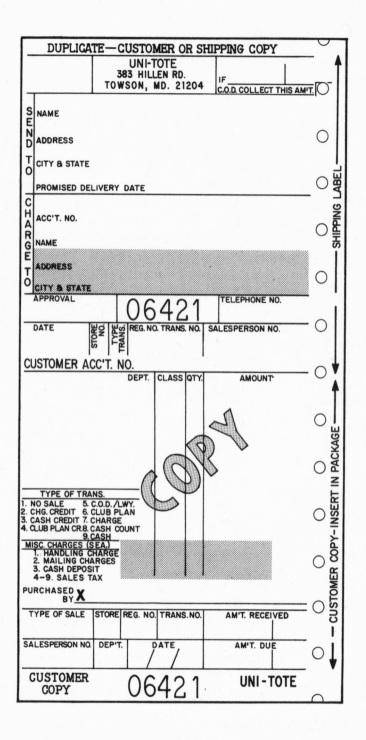

Fig. 3-3. The UNI-TOTE sales slip is automatically printed by the register and includes key-entered data as well as the automatically supplied date, store number, register number, transaction number, subtotal, and total.

forces data entry by not cycling for a transaction step if all data required for that step has not been entered. Examples of required data are transaction type and employee number for any transaction, and customer account number for charge transactions. The accuracy of such numbers as customer account numbers and stock-keeping unit numbers can be enforced by an optional procedure called "check-digit verification."

Check-digit verification is an arithmetical procedure applied to all digits of a number to verify that they have been entered correctly. A number that must be accurately entered, such as an account number, is transformed (when it is assigned) into a number that can be verified by appending a digit to the beginning or end; the value of the digit is computed by multiplying each digit of the original number by another predefined digit, which may depend on the original digit's position, and then adding the products. The check-digit value is usually the result of subtracting the least significant digit of the sum of the products from 9.

After check digits have been incorporated into each important number, the numbers can be entered into a device with a check-digit verification feature (Fig. 3-4). This device performs the same predefined arith-

Keyboard column	8	7	6	5	4	3	2	1	
Contents	CD	1	2	3	4	5	6	7	
Weight		X7	X6	X5	X4	X3	X2	X1	
Products		7	+12	+15	+16	+15	+12	+7	=84
Total subtracted from next higher multiple of 10 = check digit									90 −84 6

Fig. 3-4. Check-digit calculation for original account number 1234567, using the "Module 10" system, gives a result of 6. The check digit is incorporated as the leftmost digit for every subsequent use of the account number, for example, 61234567.

metical procedure on the original digits and then checks that the result is the same as the check digit appended to the number. If not, some sort of error indication is given. Many different check-digit verification systems exist, based on different arithmetic algorithms; most provide a high probability that digits cannot be transposed or entered as another value, but none are completely foolproof. Incorrect entry of check-digit numbers into the UNI-TOTE register causes an error indication, at which point the number can be reentered or a pseudonumber entered so that the register can continue operating.

The second aspect of UNI-TOTE system design that improves on the man/machine/machine/computer interface is its method of data collection and recording. Instead of recording transaction data on paper tape or on an OCR-font printed journal tape at each register, the system routes transaction data entered at each register to a central location via telephone lines or cable and there records the data on a magnetic tape recorder (See Fig. 3-5). The central location need not be in the store. All data entered on the register keyboard, plus the register number, transaction number, store number, date, and the register-calculated totals, is recorded on the tape. A timer, which records time data on the tape every 15 minutes, can be added at the central location so that the computer will summarize transaction data with respect to time. Tapes are processed periodically or when they are filled by removing them from the central recorder and placing them on a computer-system tape drive. Thus the handling aspects of providing computer input are much improved over previously mentioned systems, since physical handling is limited to a single computer-compatible tape. Errors due to damaged or lost punched or printed tapes and time delays due to the physical requirements of removal, collection, transportation, and loading of these kinds of tapes are eliminated.

If necessary, the UNI-TOTE centrally recorded tape can be removed at any time for computer processing, thus reducing the data processing time lag to any desired interval. It should be noted, however, that tape removal prevents data from being centrally recorded, as does any failure of the communications hookup or the magnetic-tape unit. Data will not be lost, since it is retained on the locked-in sales-slip audit copies at each register, which continue to function normally (except for check-digit verification) in case of such difficulty. A nonrecording condition is indicated at the registers by special sales-slip printing actions during each affected transaction. A backup tape unit can be added to prevent nonrecording during tape removal and single tape unit failure.

Since the only totals provided by the register are sales-slip subtotals and totals, and since the central recording unit does not develop totals, all summary information, including register sales totals for the day, must be derived from computer processing of the central magnetic tape. This could be a drawback if register, department, or store totals were required at frequent intervals, especially more than once a day; but leaving these summary tasks to the computer has the advantage of saving labor.

A third benefit introduced by the UNI-TOTE system is not related to the improvement of man/machine or machine/computer interfaces, but is the result of having a set of register terminals connected to a central location via communications facilities. That benefit is the verification of credit via register inquiry. As an option, a drum memory unit can be added to the

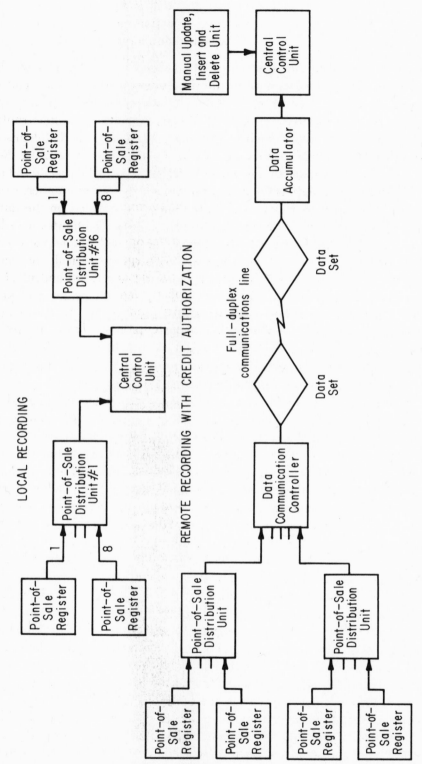

Fig. 3-5. The UNI-TOTE configuration locates the control unit and tape recorder at a control store location or at a location remote from the store.

21

central control unit; the drum contains customer account numbers of stolen cards, overcharged accounts, or any accounts requiring special attention. If such a number is entered during a transaction, the credit verification feature triggers a warning light on the entering register, signaling the salesperson to contact the credit office for instructions. Entry of forged account numbers is safeguarded by check-digit verification. This method of credit verification, called "negative credit authorization," enables the rapid checking of any charge transaction, not just those over a specified floor limit, eliminating heavy losses that can occur from fraudulent charges under the floor limit. The time-consuming procedure of telephoning credit authorization requests, except those for extraordinary amounts, is also eliminated.

The original UNI-TOTE system, even though significantly improved by UNI-TOTE's new 300 system, serves as the precursor of equipment-aided data entry and an automatic and direct method of data collection and recording. Included with these benefits is a central, communications-oriented credit-verification system. Many of the errors resulting from difficult or incomplete data entry procedures, and time delays resulting from an indirect method of data recording have been eliminated by UNI-TOTE. However, the introduction of UNI-TOTE is just the beginning of the development of a truly symbiotic retail information system. With all its power, the computer could still do much to aid the entry of data into a retail system and also to provide more immediate data processing capability than that obtained by operating on the system output through a separate system—in short, closer ties could be established between the computer and the transaction entry routed to data recording retail equipment.

4. THE ROAD TO TRUE POS

REGISTRON

In late 1967 Information Machines Corporation took another step toward a truly symbiotic retail data system by announcing its Registron point-of-sale terminal. Registron (Fig. 4-1) initially had many similarities to UNI-TOTE, but did differ in three important respects.

Like UNI-TOTE, the Registron keyboard (Fig. 4-1) has a full arrangement of numeric key columns, but instead of using the same columns to enter different types of data, all columns are arranged into specific groups, each dedicated to the entry of a specific data category. The column arrangement can be varied according to customer specification, usually ranging from 13 to 24 columns, grouped as desired. Most often, separate column groups are assigned to salesperson number, department number, style number, and price. Another column is a labeled set of keys for transaction type. Thus, a cash sale of a single item can be entered by just one line entry. Additional merchandise, and customer account number for charge transactions, require additional line entries, but certain key columns valid for the entire transaction, such as those for transaction type and salesperson number, are locked down until the transaction is completed. Thus it is not necessary to depress these keys for each line entry.

There seems to be no major advantage of Registron's keyboard arrangement over that of UNI-TOTE, or vice versa, since essentially the same number of keys must be depressed on either and the saving in number of line entries by Registron is compensated by the more compact key arrangement

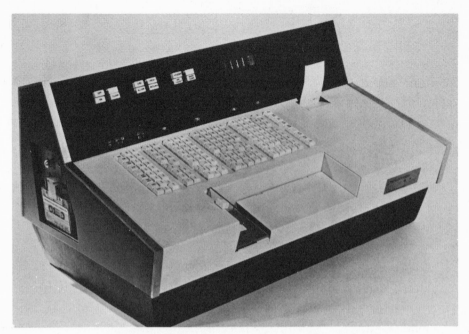

Fig. 4-1.   Registron point-of-sale register introduced local cassette
recording and merchandise tag-reading capabilities.

in UNI-TOTE. No sequential indicators are needed in Registron. However,
the register can be set to require data entry in specified columns.

The first important addition offered by Registron is its capability to
read a Dennison or Kimball punched merchandise tag. Instead of entering
department, style, and price information into the keyboard for a sales item,
the salesperson removes a punched merchandise tag containing the same
data from the item, places the tag on a special reading bed, and depresses
an entry bar over the tag; the register then reads the data encoded into the
tag instead of reading the data from the keyboard. This provision for automatic data input has obvious advantages over manual input via the keyboard: It is faster and it eliminates data transcription errors. Faster data
entry is appreciated by both the salesperson and the customer; moreover,
the manager also receives more accurate data. The introduction of automatic tag reading as part of a sales transaction is a good example of how the
man/machine interface has been modified to improve computer handling
of tasks and the quality and timeliness of data supplied. Two new options
for the Registron register are a hand held merchandise tag reader and an
embossed credit card reader, both of which will be discussed in Chapter 5.

Registron registers are usually connected to a central magnetic-tape
recorder, as in the UNI-TOTE system. However, many alternate types of data

recording are available, the most common being the use of individual magnetic-tape cassettes located in each register. The cassettes are standard C90 ⅛-inch magnetic tape cassettes (certified), which can normally record all transaction data for a complete day. Many methods are used to process the cassette data. Up to 16 cassettes can be transcribed onto a single ½-inch computer-compatible magnetic tape by an IMC translator/pooler. Small installations without their own data processing facilities mail their cartridges to a service center for processing and receive preselected reports by return mail at specified intervals. Other methods provided for data collection and recording include time-shared telephonic or direct on-line connection to a central computer, transcription of single cassettes onto punched paper tape, and direct connection to an IBM 029 key punch.

The variety of output methods of the Registron allows the machine/computer interface to match the needs of many types of retailers. Although manual handling is required for the less expensive methods, it involves only single compact cartridges, not bulky punched paper tapes, easily damaged printed tapes, or bundles of merchandise tags and sales slips. The time-shared or on-line methods are much more expensive, but eliminate manual handling operations. Since Registron did not provide on-line computer capability until recently, the discussion of this important subject is deferred to Chapter 5.

Other aspects of the Registron system are much like those of UNI-TOTE. Two-part sales slips are printed at the register and contain all data entered at the keyboard or by the automatic tag reader; one copy is retained inside the register as an audit copy. The register performs subtotaling and totaling functions as required, but all other summary data must be obtained from computer processing of the register output. (Recent versions of the register offer optional increased mathematical capabilities; see Chapter 5.) Optional credit verification on a negative basis from register inquiry is like the UNI-TOTE capability.

TRADAR

The most advanced and most exciting development in the early stages of POS systems was General Electric's TRADAR (TRANSACTION DATA RECORDER) system. The company began on-site testing of the system late in 1967 in the J.C. Penney store in Glendale, California.

The TRADAR system was significantly different from any other existing retailing system because it was entirely on line, connecting up to 1500 POS registers to a real-time processor. The processor for this system was actually a pair of GE-400 series central processors, which were connected to the registers by a pair of efficient multiline controllers and a line switch. All phases of system operation, including terminal operation, terminal/processor communications, message handling, and data recording and processing, were controlled by the processor in conjunction with its multiline controller. The dual processor arrangement enabled one processor to function while the other stood by in case of failure; the line switch could automatically switch all register lines to the standby processor and its multiline controller almost simultaneously.

In 1967 the POS registers were revolutionary in their approach to a close-working man/machine interface. Data input procedures represented significant advancements. The keyboard itself was a simple arrangement of a 10-key numeric group with about a dozen control keys and a numeric data display to show what had been entered. Once the salesperson had selected a transaction type by entering a code number into the keyboard, all further operations in a transaction were governed by a close union between the register and the processor. A multimessage display on the keyboard was illuminated by the processor at each step to instruct the salesperson. The sequence and type of step depended on the type of transaction, a great number being permitted. If the step required entry of the salesperson number, the display would illuminate a message such as SALESMAN ID; if the step required entry of item quantity, the display would so indicate. Thus, close communications were established between the salesperson and the computer. All data required by the processor was requested when needed, and the processor would not permit operations to continue until the request had been satisfied. The detailed operating instructions coupled with the simple keyboard enabled fast and practically error-free data entry.

Another development that greatly aided data entry was a system for automatically reading merchandise tags and credit cards. A reader on the front of the register could read specially encoded tags and cards, thus eliminating the keyboard entry of inventory and account number data. The tags (Fig. 4-2), called Meritags by their developer, Dennison Manufacturing Company, contain inventory data coded into a magnetic area on the back of each tag. The magnetically encoded data was scanned by the register reading device by a circular reading method. These tags usually included the same types of information found on conventional and punched merchandise tags. They were prepared by special equipment sold by Dennison.

Preparation costs were more than that of the other types of tags, but no further tag processing or handling was required other than by the salesperson, so that savings here were compensatory. Credit cards could be similarly encoded, with a magnetic area representing the customer account

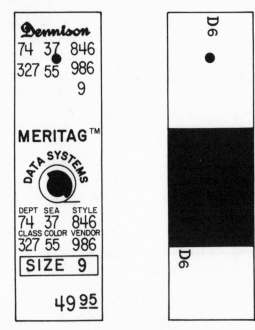

Fig. 4-2.   Dennison Meritags encode inventory data on two concentric tracks within a magnetic area on the back of the tag.

number. With this automatic reading capability, a salesperson could enter an entire charge transaction by entering a charge-transaction request on the keyboard, removing the magnetically encoded merchandise tag from each item and placing it in the reader, and then inserting the customer's credit card. No additional keyboard entry was required other than per-haps entering the salesperson's identification number and depressing the subtotal and total keys. Credit verification was automatically performed by the processor when the card was inserted into the register; normally, the system provided negative verification. The simple keyboard, automatic sequential instruction display, and the magnetic ticket reader combined to form a simple, rapid, and error-free method of transaction data entry.

Not only did TRADAR improve the man/machine interface, it also im-proved the machine/computer interface. Since TRADAR was an on-line sys-tem, the data could be processed as soon as it was entered from a register. This capability could provide many valuable types of immediate retail information. For example, the TRADAR system could maintain a current file of large inventory items in stock and immediately modify the file when one was sold. Usually, however, the processor was primarily occupied during the sales day with terminal control, negative credit checking, and data collection onto temporary magnetic tape or disk storage. At the close of business, the transaction data could be processed, files updated, and reports generated in a matter of minutes, since no manual medium-handling or transcription was required. The on-line design of the system not only permitted selective file updating and rapid processing at the close of business, but also handled requests from remote teletype terminals for file data and a limited number of retailing reports at any time during a sales day.

The sad thing about TRADAR is that it's not around any more. It seemed to be the true solution to the retailer's needs. General Electric and Penny's made a joint announcement in May 1969 that they had signed a $10 million contract, and in October 1969 Penny's point-of-sale manager presented such a stimulating report on TRADAR to a gathering of retailers in Los Angeles that they gave him a standing ovation. GE representatives, Penny's representatives, salespeople, and customers all indicated that the system was working well and that they were happy with it. However, in December 1969, Penny's abruptly announced that it was terminating the use of TRADAR, but did not say why.

It is possible that Penny's discarded TRADAR because the system was dependent on the central processor. Even though a backup processor was part of the system, certain types of failures could cause the system to become inoperative, and data entered during such a failure might be irretrievable. Such an occurrence was unlikely and no such failure had occurred in the Penney-GE trial, but the possibility that it could happen in such a large system was frightening. However, probably the most significant weakness of TRADAR was its use of two medium-scale processors for data collection and terminal control. Such a design required an extraordinarily large system to reduce the per-terminal price to a reasonable level. The capacity of TRADAR was 1500 terminals, with 400 terminals the smallest efficient size. A 400-terminal system would have cost approximately $3.5 million. Besides being cost-effective only for large systems, TRADAR also required changeover to a different type of merchandise tag and credit card in order to utilize all the benefits of the system.

These difficulties might not seem insurmountable for a large retailer like Penny's, but at the time of the announcement two things probably precipitated Penny's decision. First, both GE and Penny's had replaced the top men who were handling TRADAR; second, hints of better retail systems to come were circulating among retailers along with the two releases of POS systems, which did not require a TRADAR-like commitment. Therefore it appears that Penny's adopted an attitude of wait and see rather than do and possibly die. Thus ended a true breakthrough in retail information systems.

RICCA

At the time of TRADAR, another company had also been working on an on-line POS system and had scheduled its release for 1969. The company, Ricca Data Systems, was assuredly watching TRADAR, and about the time of TRADAR's demise, changed its mind about releasing its POS system. Instead, it concentrated its efforts on an off-line inventory-recording device, which it released in 1970. This device was a portable, battery-powered tag reader that collected inventory data onto magnetic-tape cassettes. Ricca's on-line device was finally released in early 1971. Since it is significantly different from TRADAR, there should have been little cause for worry about the system's chances of success, because of what had happened to TRADAR.

Ricca's POS system is a network of remote POS registers (Fig. 4-3) controlled by a central processor, but it is on a much smaller scale than TRADAR.

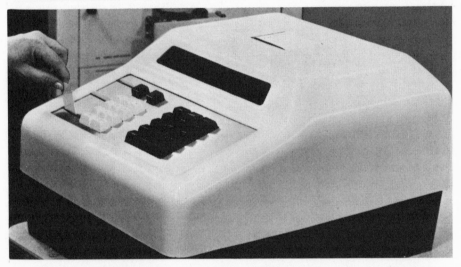

Fig. 4-3.   Ricca MARS on-line point-of-sale system controls register functions at a central minicomputer.

The central processor is actually a minicomputer, which is responsible for system control and data transfer between the POS registers and a disk storage unit. System capacity is 100 terminals, which are interfaced to the control minicomputer via a multiplexor. The registers read merchandise tags, the old familiar Dennison and Kimball punched-hole variety; they also read credit cards and salesperson identification cards punched in Dennison or Kimball code. Despite its smaller size, the Ricca system offers most of the advantages that were provided by TRADAR.

The disk files of the system are accessed by the control minicomputer and are used to maintain inventory and customer account files. These files can be checked and updated for inventory flow and credit verification during each transaction. The credit verification feature is provided in three configurations: a negative system, described previously; a partial positive system, which assigns each customer account a special allowance that is updated for each transaction; or a full positive system, which maintains and updates current customer-account balances.

Both inventory and credit information on the disk files can be accessed by the store's central computer system for transaction data processing, or the files can be formatted by the POS system minicomputer for limited report output to punched paper tape, magnetic tape, or a data communications system.

These features of the Ricca are the outstanding advantages provided by an on-line system; yet this system requires only a minicomputer, not a medium-scale computer, for system control functions. However, a system with even 100 registers seemed too large for many retailers. What they wanted, and what many manufacturers were developing, was a smaller system that could be expanded as required. Such a system is discussed in Chapter 5.

5. THE NEW SYSTEMS

Point-of-sale systems became really prominent in 1970 and 1971 with releases by six manufacturers and system revisions by two others. Each system worked toward providing both a symbiotic device and a device that would solve many of the retailer's problems of transition from a conventional cash register system to one that was computer-oriented only. The new systems introduced many innovations in merchandise tag coding and reading, in methods of salesperson-register interaction, and in a variety of system configurations. The period was evidence that EDP equipment manufacturers had indeed something valuable to offer the retailer.

SINGER'S MDTS

The first official release of a point-of-sale (POS) system in this period was the Modular Data Transaction System (MDTS) by Singer's Business Systems Division. Primary among the innovations introduced by this system is its capability of being configured in a number of ways. What makes modification of the configuration possible is the design of the register terminal. Each register is basically a stand-alone unit, which means that it operates independently of any central processor. All functions necessary to operate the terminal are contained within the terminal, which effectively contains its own microprocessor, utilizing MOS/LSI circuit technology.

Before describing what the terminal can do, it is best to explain how these free-standing terminals can be part of a data collecting retail information system. Three basic system configurations are possible: individual store and forward, collective store and forward, and on line.

Individual Store and Forward System

All transaction data is recorded on a continuous magnetic-tape loop (capacity: 50,000 characters), which is contained in a special individual store and forward module (Fig. 5-1) located near the register. Each module is connected to a computer via a modem and telephone line. The computer periodically polls each module to request transmission of data that has been recorded on the tape, but such a request does not affect the module's recording abilities because it can record new data simultaneously with the

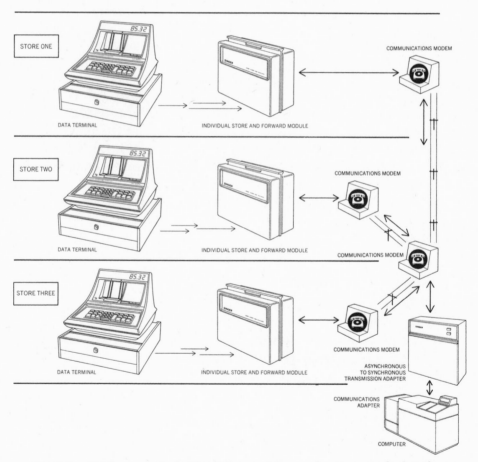

Fig. 5-1.   Friden MDTS individual store and forward system sends data from each register to its own tape loop module, which is polled as required by a central computer.

reading and transmission of old data. Automatic retransmission is performed upon error detection by the computer. The store and forward module is portable, as is the register terminal, and can share a single modem with several other modules for output communication to the computer.

Collective Store and Forward System

All transaction data is sent to an in-store System 10 processor (Fig. 5-2), which functions as a scanner, buffer, and line concentrator for up to 180 terminals. Data is formatted by the processor and stored on a Friden Model 45 magnetic-tape drive, or a Model 40 disk drive. The disk drive permits negative credit authorization upon terminal request. The processor is

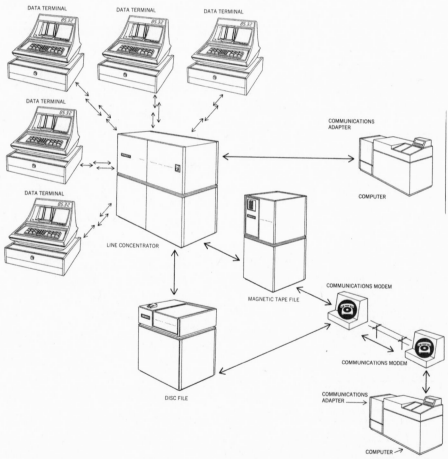

Fig. 5-2. Friden MDTS collective store and forward system sends data from each register to a central tape unit, which is accessed as required by a central computer.

connected to a computer via modems and a telephone line, and is polled typically at the end of each sales day to transmit the data stored on the disk or tape to the computer. Using the processor link, the computer can modify the information stored on the credit file. An alternate method of collective recording connects up to 24 MDTS registers to a Singer 4301 Data Recorder, which records sales data on computer-compatible magnetic tape, but which does not forward the recorded data to a computer.

On-Line System

All transaction data is sent via an in-store System 10 processor (Fig. 5-3) and high-speed communications facilities directly to a remote on-line computer such as the IBM 360/30. The on-line computer can provide positive

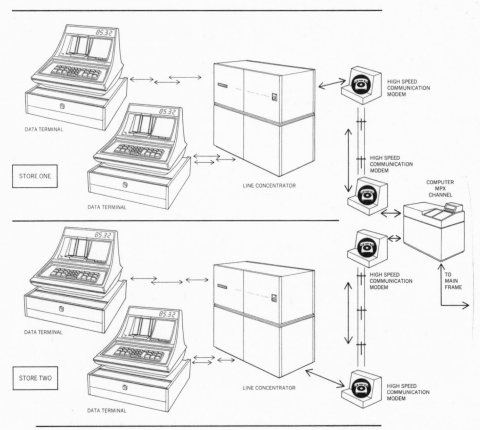

Fig. 5-3.   Friden MDTS on-line, real-time system provides positive credit verification and immediate transaction processing.

credit verification and instant file maintenance and transaction data processing.

As mentioned above, each register terminal (Fig. 5-4) is free standing and has its own microprocessor. The microprocessor stores a predefined operating sequence for entering a transaction. The sequence does much to aid data entry. First, it causes the illumination of function keys on the keyboard to indicate which keys are valid at a certain transaction step, and sets interlocks to prevent depression of others. Second, it causes the necessary arithmetic calculations to be performed as they are needed in the transaction. These calculations include line extensions (multiplying item quantity by unit price to determine amount), subtotals, taxes, discounts (multiplying subtotal or item amount by stored percentage), totals, change

Fig. 5-4. Friden MDTS data terminal contains its own programmable microprocessor to control all register functions. Proper operating procedure for specific transactions is indicated by the sequential illumination of control keys.

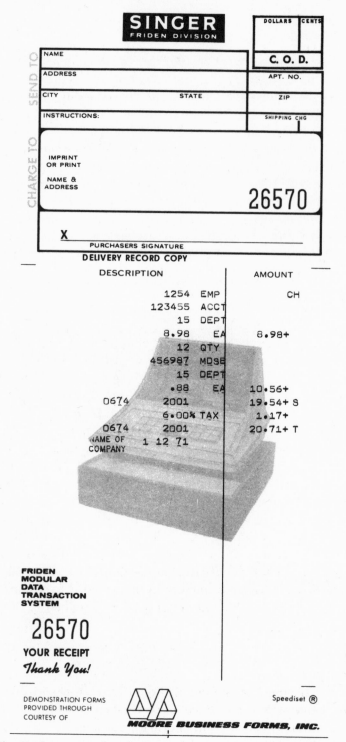

SINGER
FRIDEN DIVISION

DOLLARS	CENTS

C. O. D.

NAME

ADDRESS   APT. NO.

CITY   STATE   ZIP

INSTRUCTIONS: SHIPPING CHG

SEND TO

CHARGE TO

IMPRINT
OR PRINT

NAME &
ADDRESS

26570

X _____
PURCHASERS SIGNATURE
DELIVERY RECORD COPY

DESCRIPTION   AMOUNT

```
          1254   EMP            CH
        123455   ACCT
            15   DEPT
          8.98     EA        8.98+
            12   QTY
        456987   MDSE
            15   DEPT
           .88     EA       10.56+
  0674     2001            19.54+ S
          6.00% TAX         1.17+
  0674     2001            20.71+ T
 NAME OF   1 12 71
 COMPANY
```

**FRIDEN
MODULAR
DATA
TRANSACTION
SYSTEM**

26570

YOUR RECEIPT

Thank You!

DEMONSTRATION FORMS
PROVIDED THROUGH
COURTESY OF

Speediset ®

MOORE BUSINESS FORMS, INC.

Fig. 5-5.   Friden MDTS sales check is inserted into the front of the register for all noncash transactions. Product descriptions are numeric; individual lines are identified alphanumerically.

due, and check-digit verification on preselected numbers. The sequence can be conveniently changed to meet any retailer's requirements.

Besides performing the calculations and illuminating the lamps ordered by the stored sequence, the microprocessor prints receipts and tapes, displays data, controls communications to the individual store and forward module or to the System 10 processor, and accumulates four separate totals (cash, sales, sales tax, and discount), which can be read or cleared at any time by the use of a special access key. The sales receipt can be a tear-off type, or it can be a multipart sales check (Fig. 5-5) fed into the register from the front. Both the receipt and a locked-in audit journal tape contain all data entered plus the results of calculations, each line identified by a one-word description such as ACCT, MDSE, EA, TAX, and PYMT. Merchandise items are identified by SKU or style numbers. All data entered into the register or calculated is also displayed on a 13-digit electronic display; optionally, a separate 7-digit customer display can be used to show just the results of monetary calculations.

One further capability of the MDTS terminal is its ability to read machine/readable merchandise tags and credit cards. Friden states that it can add almost any type of OEM reader currently available, including those for Kimball and Dennison punched tags, Dennison magnetic Meritags, and others to be discussed later in this chapter.

Thus, the Friden MDTS marked the introduction of a new approach to retailing information systems. The terminals provide almost every aid possible to the salesperson, including telling him what to do next, preventing him from doing the wrong thing, doing his calculations for him, automatically reading in merchandise data and customer account numbers, performing credit verification, and preparing his receipts. All the necessary logic for these tasks (except credit verification) is contained in the terminal. Also, the modularity of the system permits any type of retailer to buy the system he needs and modify it as required; thus no large-scale commitment is required. This same philosophy is extended in some areas and diminished in others by the new systems to be described in this chapter.

NCR 280

The announcement of the NCR 280 retail system in mid-1970 was important to the growth of POS, since it was the first commitment of a major cash register to that type of retail information system.

The NCR 280 contains many of the innovations introduced by Friden's MDTS; although it is also a free-standing terminal with its own microprocessor, it differs from the MDTS in several features.

Primary among the NCR 280 innovations is its color-coded medium data entry system. This system is based on a new method introduced by NCR, which represents binary digits as an array of printed green, black, and white stripes. The colored stripes are printed on merchandise tags (Fig. 5-6)

Fig. 5-6.   The NCR color-coded merchandise tags encode inventory data with white, green, and black bars, permitting rapid data entry by scanning the tags with a handheld optical wand.

to encode inventory data, on credit cards to encode customer account numbers, and on salespersons' badges to represent employee numbers. Tags are encoded by a special NCR tag printer, which can produce many varieties of tags. One of the larger tags can encode up to 56 numeric digits, comprising such data as vendor, style, color, size, department or class, SKU, season, price, and multiple prices (3/$1.00) in the form of unit price ($1.00) and multiple-unit (3) price. Inventory data is also printed as characters on the tags; the printing characters include 19 symbols as well as the 10 numeric digits.

Tags are read by the salesperson with a handheld optical scanner (Fig. 5-7), which is attached to the register by a flexible cord composed of fiberoptics bundles. To read the tags, the salesperson passes the tip of the scanner, shaped much like a ball-point pen, over the colored bands, starting from either end and at any speed. Checks built into the register ensure a correct scan by safeguards built into the coding system; a pleasant tone sounds for a correct scan, and a harsh blat sounds for an incorrect scan. This method of tag reading permits the tags to remain on the merchandise, since only the tag surface need be accessed by an easily handled device; NCR provides a variety of pin-on, stapled, and adhesive tags. The NCR colorcoded tag and card coding and scanning system is the fastest method yet devised to enter inventory data and customer account and salesperson numbers into a retail transaction system. Three other systems similar to NCR, and equally fast, are discussed later in this chapter.

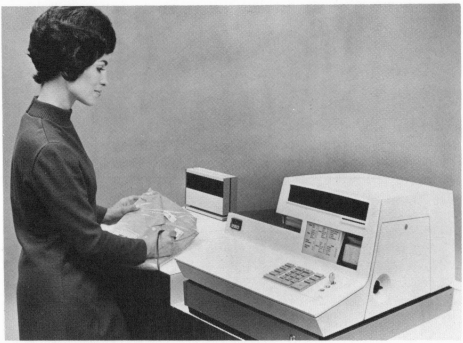

Fig. 5-7. The NCR 280 register simplifies data entry by a combination of a small 20-key keyboard, a transaction code chart that illuminates only when needed, a selectively illuminated message display panel, and an optical wand for reading tags, cards, and badges.

Simplicity of data entry is also carried into the arrangement of the keyboard, which comprises 10 numeric keys and just 10 function keys. Only five of the function keys are used in a normal transaction: ENTER, NONTAX (to prevent tax calculation for nontaxable items), SUBTOTAL, CURRENT TOTAL, and TOTAL END TRANS. The other five are for clearing entries, correcting errors, and manually entering prices.

The ability to perform a complete set of transaction types, despite the simple keyboard, and other retail functions, however, is retained, and is based in a component called the operator qualification panel. This panel, located on the front of the register, is simply a chart that lists all possible types of transactions and other functions the register can perform. Associated with each transaction type or function is a one- to three-digit number. When the register is ready to enter a new transaction, the panel is illuminated, and the salesperson enters into the keyboard the numeric code of the transaction type he wants. The entry of this code causes a preprogrammed sequence to be executed in the microprocessor, which controls all operations of the register until the end of the transaction. Instead of illuminating valid function keys, the sequence causes instructions in a message display panel to be illuminated at each transaction step, to indicate what kind of data should be entered next.

The microprocessor, under control of the sequence, also performs the same types of functions performed by Friden's MDTS microprocessor, including arithmetic calculations, receipt and tape printing, and data display. Tax calculations, which can be fully or partially automatic, are performed by using stored percentages or stored break-point tables. If further sequence selection is required during the course of a transaction, such as selection of a type of fee to be charged, the qualification panel lights up so the salesperson can select the proper fee and enter its numeric code; the panel then darkens and the transaction continues. Stored transaction sequences can be changed at any time from the keyboard, using a special access key.

The printing device of the NCR 280 is unique in that it contains three separate printers for the audit journal tape, the sales check, and the tear-off sales receipt. A multicopy sales check must be inserted into the register at the start of any transaction, except cash-take; otherwise, a safeguard prevents operation. Each printer formats the printout to suit its function.

Unlike the MDTS, the NCR 280 has one standard system output, a computer-compatible magnetic tape. Up to 48 POS registers can be interfaced to a data collector and recorder, which accepts data from each register at the end of each transaction (or when the register's data buffer is filled) and then records the data on magnetic tape. To reduce expenses for small systems, the data collector and recorder can also be obtained with a maximum capacity of 32, 16, or 8 POS registers. If, for some reason, data cannot be recorded on the magnetic tape, the terminals are so notified and then allowed to continue operation; they perform all functions normally except that each terminal's microprocessor adds special marks to the audit journal tapes to indicate that data has not been transmitted and recorded. Cash totals are maintained on the journal tape to facilitate reentry of the data when the system becomes operative.

Two important options for the 280 system are credit verification and configuration as a collective store-and-forward system. The latter option permits data collector and recorders to be polled for data transfer to a central computer.

Sweda Series 700

A system with many basic similarities to the Friden and NCR systems, but with many innovations of its own, is the Sweda Series 700 retail system, announced in March 1971. The system can be supplied in three configurations, but not all configurations use the same register terminal: local and central recording; and on-line, in-store.

1. *Local Recording System.* Free-standing registers, each with its own programmable microprocessor, record transaction data on their own internal magnetic-tape cassettes. Cassettes are sent or transmitted to a pooler for transcription to computer-compatible magnetic tape.

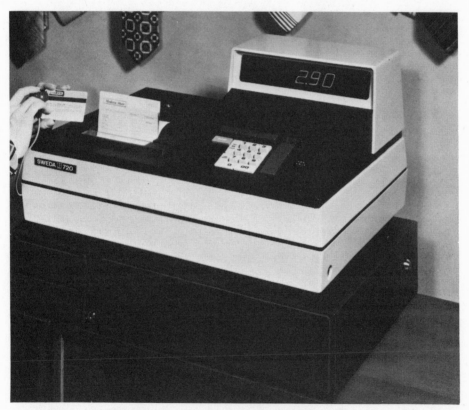

Fig. 5-8.   Sweda System 700 point-of-sale register is characterized by its variable-function control keys and its ability to read magnetically encoded cards, badges, and merchandise tags with a handheld scanner.

2. *Central Recording System*. The free-standing, internal microprocessor-controlled registers are equipped with data communications capability to enable transaction data recording on a central computer-compatible magnetic-tape unit and negative credit verification functions.

3. *On-Line, In-Store System*. From 16 to 256 register terminals are controlled by a single minicomputer. Transaction data is recorded on a central computer-compatible magnetic-tape unit, but a disk file unit permits positive credit verification from any terminal and limited real-time preparation of sales volume reports upon request from a special console. More complicated reports are prepared off-line.

The register terminal (Fig. 5-8) used for any of the configurations has

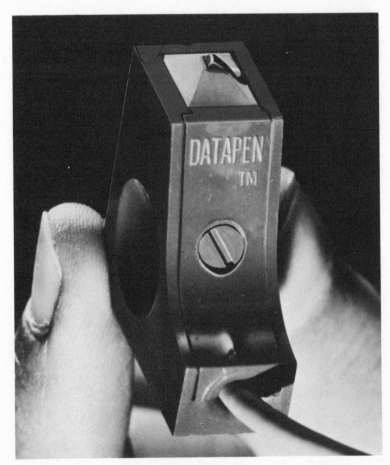

Fig. 5-9.   Sweda Datapen reader scans magnetically encoded tags on any type of merchandise and magnetic stripes on credit cards and employee badges. A light in the top of the reader and an audible signal indicate a valid scan.

a keyboard that is really a combination of those found on the Friden and NCR registers. Sequentially illuminated instruction lamps are positionally associated with six function keys, which change their function according to the contents of the instruction messages being displayed near the individual keys. The digital readout and receipt/sales check/audit journal printer is basically like that on the MDTS.

The most significant feature of the Sweda 700 is its medium encoding/reading system, not because of the reading method, but because of the principle used in encoding the merchandise tags. The system uses a special dual-language tag produced for the Sweda 700, which is the familiar Kimball punched tag with magnetic stripes added. Inventory data is encoded redundantly as printed numeric characters, punched holes, and a magnetic record.

The Sweda 700 register reads the tags without removing them by scanning the magnetic stripe area with a compact, handheld, magnetic scanner (Fig. 5-9), passing its tip longitudinally along the strip in either direction. The purpose of using a conventional Kimball punched tag (Fig. 5-10) as the medium for the magnetic stripe is to ease the retailer's transition from a punched tag system to the faster magnetic-tag system provided by the 700. This orderly tag system changeover is very valuable to many retailers. Magnetic-stripe encoding is also used for customer account numbers on credit cards and employee numbers on employee badges.

Fig. 5-10. Kimball memory tag encodes inventory data in two machine languages: one the traditional punched hole code for reading by conventional tag equipment, and the other a magnetic code for reading by magnetic scanners such as Sweda's Datapen.

Pitney-Bowes-Alpex Spice

There are also several new retail information systems that differ significantly from the Friden, NCR, and Sweda systems in that they do not comprise a collection of free-standing terminals. One of the most flexible of these is the Pitney-Bowes-Alpex SPICE (sales point information computing equipment) system developed by a joint effort of Pitney-Bowes, Inc., and Alpex Computer Corporation, and announced in 1969. In the standard configuration of SPICE, the POS registers are connected to an in-store dual controller comprising an active and a standby minicomputer and interface circuitry. The controller governs most aspects of POS register operation, including transaction sequence control, instruction message illumination, arithmetic calculations, and error checking. Capacity of the basic dual controller is 32 registers; in its most elementary configuration it can accumulate up to 160 separate totals. If desired, a SPICE register can be configured as a free-standing, local-recording unit.

A dual-cassette magnetic-tape recorder can be added to the dual controller to record transaction data; the tape is written with a time mark at 15-minute intervals so that sales can be analyzed on the basis of time. Recorded data is transferred to a computer center by mailing the cassettes or transmitting the data via telephone lines when polling is requested by a central automatic-calling unit and communications controller. Once the data has been received at the computer center, it can be recorded on computer-compatible magnetic tape or used immediately to update a central large-item inventory file and central credit file, both of which can be accessed for current data by terminal request. Within the store, a negative credit file can be included for credit verification, and a device called a store-price memory can be added to maintain the current price of any inventory item. This price is supplied to any register once that the item SKU number has been entered into the register by a salesperson. Price data can be modified from any terminal at any time. Such a central price system guarantees accurate pricing and makes price changing a fast, simple operation.

The appearance of the SPICE register keyboard is unique, but its operating characteristics are very similar to those of the new systems already discussed. Keys are arranged as a vertical column of 9 quantity keys, a separate set of 10 numeric keys for other numeric data entry, and a varying number (around 13) of function keys placed according to type. But, as in the other systems, messages to guide the salesperson are sequentially illuminated, data is displayed on a digital readout, and a numeric tear-off sales receipt is printed, along with a locked-in audit journal tape.

The SPICE system provides an optional merchandise tag coding and scanning system much like the NCR color-coded system. An optical scanner (Fig. 5-11) called PEPPER (photoelectric portable probe/reader) scans merchandise tags encoded with variably spaced, parallel black bars (Fig. 5-12) instead of colored stripes. This scanner operates like the one on the NCR 280. Although tags cannot be encoded with the same versatility and

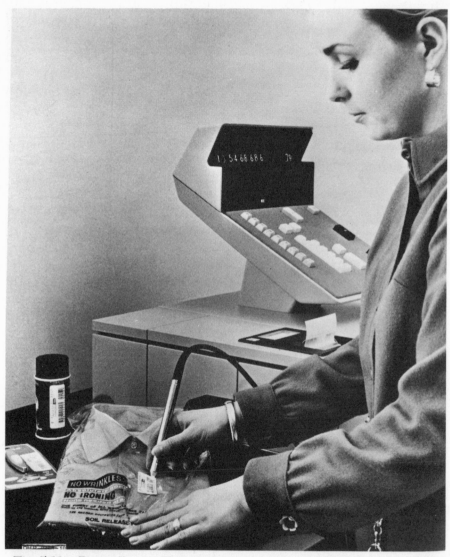

Fig. 5-11. Pitney-Bowes-Alpex SPICE register can scan special merchandise tags with an optical scanner called PEPPER. The unique keyboard arrangement may be modified to suit a retailer's requirements.

Fig. 5-12.   Pitney-Bowes-Alpex merchandise tags encode inventory data by the density and spacing of black parallel bars for reading with a handheld optical scanner.

do not have the capacity of NCR tags, the tag preparation equipment is simpler and less expensive. Besides PEPPER, there is another register option, a built-in credit card reader.

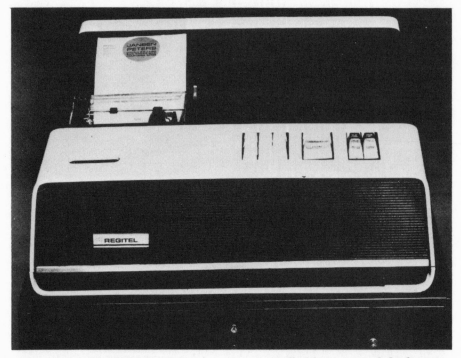

Fig. 5-13.   American Regitel terminal is controlled by a central dual mini-computer, which accesses disk files for alphanumeric customer name and address and product descriptions.

Regitel

In July 1970 the American Regitel Corporation delivered a POS register system like the SPICE system. The Regitel system utilizes a central dual

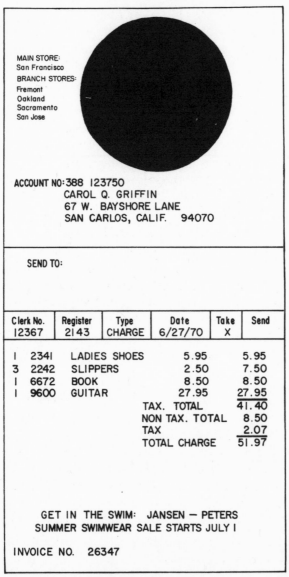

MAIN STORE:
San Francisco
BRANCH STORES:
Fremont
Oakland
Sacramento
San Jose

ACCOUNT NO: 388 123750
      CAROL Q. GRIFFIN
      67 W. BAYSHORE LANE
      SAN CARLOS, CALIF.   94070

SEND TO:

Clerk No. 12367	Register 2143	Type CHARGE	Date 6/27/70	Take X	Send

I	2341	LADIES SHOES	5.95	5.95
3	2242	SLIPPERS	2.50	7.50
I	6672	BOOK	8.50	8.50
I	9600	GUITAR	27.95	27.95

                    TAX. TOTAL   41.40
                    NON TAX. TOTAL   8.50
                    TAX   2.07
                    TOTAL CHARGE   51.97

      GET IN THE SWIM: JANSEN — PETERS
      SUMMER SWIMWEAR SALE STARTS JULY I

INVOICE NO.   26347

Fig. 5-14.   Sales check printed by the Regitel terminal contains the customer's name and address (printed if the entered account number passes credit verification) and product descriptions (printed after the product identification code is entered).

minicomputer for logic, control, and communications functions. The minicomputer has a capacity of 40 POS terminals and uses disk drives to provide an unusual feature, the printing of fully descriptive sales checks
(Fig. 5-14). The drives maintain two principal files, an inventory file containing alphanumeric product descriptions keyed to product identification codes, and a credit file with customer names and addresses keyed to
customer account numbers. This latter file can also store flags on certain
accounts for negative credit verification.

Several features of the Regitel system are similar to those of other new
POS systems. The system is described by the following list of occurrences
during a typical charge transaction:

1. The salesperson enters his employee number via the keyboard or a
badge/card reader (any OEM reader can be supplied); he then enters the
transaction type, causing the transaction type to print on the sales check
and an instruction lamp illumination sequence to be initiated.

2. The salesperson enters the customer account number via the keyboard or the badge/card reader. A negative credit check is performed at
the central minicomputer; if the credit check passes, the customer's name
and address are automatically printed on the sales check.

3. The merchandise identification numbers, prices, and quantities are
entered via the keyboard or a merchandise tag reader; alphanumeric product descriptions retrieved from disk files are automatically printed on the
sales check, and price extensions are calculated by the central minicomputer and also automatically printed.

4. The salesperson requests subtotal, tax, and total, which are calculated by the central minicomputer and printed on the sales check with an
alphanumeric description.

The ability to print a truly human-readable sales check with full alphanumeric descriptions is important because it provides a sales transaction
document that can be easily understood by anyone.

The register prints and stores an internal audit tape containing the
same data as the sales check. The portions of this data that were entered
into the register by keyboard or reader or which resulted from calculations
are normally routed by the computer to a computer-compatible magnetic-
tape unit. Transaction data can also be sent to remote devices for special
applications; for example, shipping labels for SEND transactions can be
printed in real time at a warehouse.

Olivetti TC 600

Olivetti, the company that built the production-model terminals for

GE's TRADAR system, has developed its own, smaller version of TRADAR, the TC 600. In this on-line system, up to 128 POS registers are connected to a concentrator via a multiplexor, and the concentrator is connected on line to a computer (Fig. 5-15). The concentrator is a minicomputer that

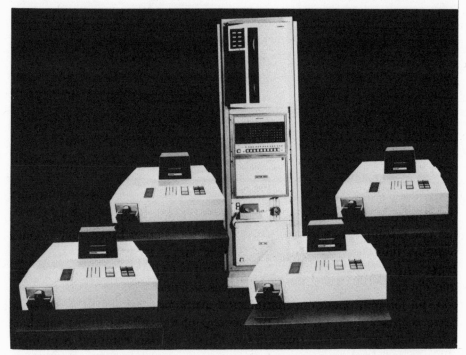

Fig. 5-15.   Olivetti TC 600 terminals are connected to a minicomputer controller/communications processor, which is interfaced to a magnetic-tape unit or an on-line computer.

functions as a terminal controller and communications interface, while the computer collects and processes data sent from the terminals. Olivetti's system can also be configured for off-line collection by substituting a magnetic-tape transport for the on-line computer.

Transaction Systems

An unexpected step in the simplification of the operator/machine interface came in 1971 when Transaction Systems, Inc., designed a retail information system that eliminated the keyboard. Instead of using numeric keys and control keys to enter data into the register, the system uses a cable-connected magnetic scanner to scan magnetically encoded paper strips. The magnetic scanner (Fig. 5-16) is about the same size as Sweda's scanner and operates in much the same way. The magnetic strips are used not

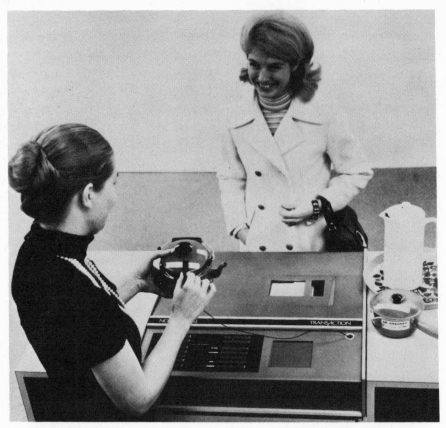

Fig. 5-16.   Transaction Systems point-of-sale register eliminates the keyboard, accomplishing data entry and control functions via hand-held magnetic scanner.

only to encode inventory data (Fig. 5-17) on adhesive merchandise labels (numeric product code, alphanumeric description, price, and a taxable/nontaxable indicator), customer account numbers on credit cards, and salesperson identification numbers on employee badges, but also to encode

──────TRANSACTION   SYSTEMS──────
SKU   ITEM   PRICE
351 951   TEA POT $ 6.50

Fig. 5-17.   Transaction Systems merchandise tags are magnetically encoded with SKU number, alphanumeric description, price, and a taxable indicator; they can be affixed to any type of merchandise.

control commands. Several magnetic strips located on the register's front panel represent individual control functions such as subtotal, tax, cash, charge, COD, layaway, and total; these are traced with the magnetic scanner to enter selected control functions.

Each control strip is accompanied by a message lamp. The message lamps of valid control strips and operating instruction messages are selectively illuminated during a transaction by a programmed operating-sequence resident in the register's microprocessor. The operating sequence also safeguards against wrong types of entries. Exception type entries, such as an account number for a customer who has forgotten his card or a variable discount rate, can be entered into the register via an auxiliary 10-key numeric keyboard hidden behind a sliding door. Register operation is complemented by printing sales slips with alphanumeric product descriptions.

In a standard system configuration, the POS registers are connected to a dual minicomputer that performs the following tasks:

1. Polls the registers and initiates transmission of transaction messages.

2. Performs check-digit verification on account, employee, and SKU numbers.

3. Forwards credit card numbers to the customer's credit-verification system and returns results.

4. Records completed transactions on computer-compatible magnetic tape, adding time of day.

```
        10/20/70              PROMOTIONAL FLASH
```

DEPT	VENDOR/ CLASS SKU		DESCRIPTION	QTY	SALES	PRICE
1	1.	4172	REG 14.99 MIXER	14	139.86	9.99
1	2	----	COFFEE MAKERS	97	857.17	
27	1	8050	WHITE STAG SHIRT	12	143.88	
76	---	----		417	1275.17	
20	5	917	POLAROID	39	1186.50	

Fig. 5-18.   Transaction Systems point-of-sale system includes software for all analysis and reporting functions, including the generation of periodic or on-demand promotional flashes that (in order of listing) verify price markdowns and list sales by class, by SKU, by department, and by vendor.

5. Accumulates running totals on sales by store, department, and register.

6. Maintains running totals on number of items sold for 50 preselected SKU numbers.

7. Prepares requested reports (Fig. 5-18), upon inquiry or at pre-established intervals, involving totals accumulated in (5) and (6) via Teletype printout.

8. Prepares system status reports, signals stolen card use, and transmits salesperson alerts to security, all during real time.

9. Prepares detailed and summary reports at the end of the sales day from recorded data, including dollar sales and profits by division and department; gross sales, percentage of returns, number of transactions, and gross profits by division and department; audit report and cash count; and salesperson analysis and activity by hour.

All software for system operation control, including that for real-time and end-of-day reports, is provided with the system.

The task of transaction data entry is so easy and fast via the man/ machine interface of this system that Transaction System claims a complete charge transaction, including customer signature, can be completed within 20 seconds.

CHANGES TO THE EARLY SYSTEMS

The advances made by these new POS systems have affected manufacturers of earlier systems. Information Machines' Registron now has options for a cable-connected, handheld reader to scan punched merchandise tags; a credit card reader to scan OCR font-embossed credit cards; a calculating unit to perform automatic line extensions, tax calculations, and master totaling; and on-line interface capability. An advanced version of the UNI-TOTE original system has a 10-key keyboard, the ability to add a merchandise tag and/or credit card reader, and increased terminal calculating ability, including multiply.

Despite all the differences in the recent offerings of retail systems equipment, the manufacturers have agreed on two common principles. First, registers should pass on the benefits of a computer to the salesperson—by guiding operation, performing arithmetic, preparing sales slips, and checking credit. Second, retail systems should get data to the computer with minimal human intervention and as fast as possible, and provide a simple, rapid means to access, to analyze, and to report it.

The advantages of computer-oriented transaction processing systems have also been extended into other areas, as is described in Chapter 6.

6. SPECIALIZED POS SYSTEMS

The development of fast and efficient retail information systems has not been limited to retail merchandising. Many recent systems designed under the same impetus as the systems already discussed have been oriented toward the food industry. Three such systems are described in this chapter.

COMTAR

The first system, called ComTaR, is designed specifically for applications with a small number of product types, but needs improvement of its sales procedure and record data for computer processing. Information Technology, Inc., designed this system, which is currently in use in fast-service food stores (hamburger stands) and is also being directed at such establishments as cleaners and taverns.

The ComTaR's keyboard is principally composed of a column of quantity keys and a set of product-type keys (e.g., hamburger, cheeseburger, special, and large chocolate shake). Purchases are entered into the cash register by depressing the appropriate quantity/product key pair for each item type purchased. After the total key has been depressed, all entered quantity/product pairs are recalled from register storage along with a pre-set and prerecorded price for each product. Multiple-item prices are automatically totaled and then all price extensions are added to determine the sale subtotal; the tax is then calculated and added and the total is calculated and displayed at the register. As the arithmetic calculations are being performed, a sales slip is automatically printed; it lists the quantity/product pairs, subtotal, tax, total, and transaction-sequence number. This same data may also be automatically recorded on a cable-connected paper tape or magnetic-tape cassette recorder. All automatic operations after entry of the quantity/product pairs and depression of the total key are normally accomplished within three seconds.

The prestored prices and calculating ability of the register speed up food sales tremendously and at the same time provide a machine-readable record of transactions so that management gains an analysis tool to run the business more efficiently.

TRACOM

Another type of fast-service food POS register is TraCom (transaction communicator), developed by Honeywell. The register handles data entry with quantity and product keys much like those of ComTaR, but the register is connected to a minicomputer (Honeywell H112), which captures transaction data, stores it in core memory, and performs limited processing and reporting functions. Besides performing calculations for food sales and preparing reports, the registers can also be used to record employee attendance and to handle management data input and retrieval operations. At any terminal a manager, using a special access key, may request that any of the following reports be generated by the minicomputer and listed on a printer attached to the register:

1. Hourly sales.
2. Labor information.
3. Cash, volume, and item usage.
4. Comprehensive management report (punched on paper tape).
5. Errata and special instruction messages.

Any report except the comprehensive management report can be requested at any time during the sales day without affecting normal operation of any other register in the system.

The POS registers also provide for the following types of management inputs:

1. Menu item price changes.
2. Clock and day changes.
3. Employee meals and waste.
4. Voids.

The TraCom system thus offers the transaction simplification and data recording capability of the ComTaR register, and also gives management a means to access certain operating data in real time as it is required. The convenience of immediate reporting is obvious.

Other POS registers have also been designed to replace supermarket check-out registers by specialized terminals that communicate with a central computer system. Inventory Management Service has developed such a system using Honeywell equipment. With the standard configuration, POS terminals are connected via an in-store multiplexor and phone lines to a Honeywell H316 computer located at a data center.

Check-out terminal keyboards comprise quantity keys, product code keys, and color-coded department keys. The checker operates the terminal by depressing a specific quantity key, entering the numeric product code as identified by a label attached to the item, and then depressing the color-coded department key identified by the label color. (The grocery department and one-item quantity are assumed by the register if other department or quantity keys are not depressed.)

The terminal sends all register-entered data to the H316, which looks up prestored item prices; adds taxable item (as internally identified) prices and looks up or calculates tax, subtotals, and total; and computes change, coupon credit, bottle refunds, and stamps due. Odd-weight items are weighed on a scale at the terminal; the weight data is sent to the H316, which calculates the weight-based price. All prices and calculation results returned by the H316 are displayed at the entering terminal and printed on the customer's tape receipt.

Special terminals, called management interrogation devices (MID), can function as check-out terminals and as inquiry terminals to request inventory of any product, total sales by department or check stand, total store sales, value of submitted coupons, taxes paid, and other data; the special terminals can also be used to enter price changes.

In addition, an H316 computer can be placed on the store premises as a backup computer in the event of telephone line or central computer failure. The store system can be powered by batteries during a power failure.

As a future development, the registers will be equipped with cable-connected scanners so that checkers can enter product and department codes without keyboard entry by scanning optically coded labels attached to the products. The effect of such a system on supermarket check-out, especially with an automatic scanner, can be tremendous. The effect on check-out procedures, together with control and analysis features for management can radically change the supermarket business.

7. PROSPECTS FOR POS

In mid-1971 many retailers began to make major commitments to POS retail merchandising systems. For example, in June 1971 Sears and Singer completed a contract estimated between $50 and $100 million. The leader up to that time had been UNI-TOTE, with over 3000 terminals installed by early 1971.

Why did POS systems begin to appear so rapidly in 1970 and 1971, and why did retailers finally begin to accept them? One of the most significant answers to the first question was that electronics technology had finally adopted extensive use of MOS/LSI circuitry. Thus, all functions for a true POS device could be assembled into a compact, low-cost package. Witness the rapid development at the same time of complex programmable calculators, handheld calculators, and minicomputers.

Two important answers to the question about retailers' acceptance are that a significant number of test marketings for POS systems had been completed or were in progress at that time and that retailers could finally see a real need for the systems. Credit verification was becoming an absolute necessity and many retailers installed one of the dedicated credit authorization systems offered by Digital Data Systems, TRW Data Systems, Data Source Corporation, Credit Systems, Pitney-Bowes, and Concord Computing Corporation. The success and power of these credit authorization systems have often served as a model of what a POS system (adding all the benefits described earlier) can do for a retailer willing to change his methods.

It appears that many advances can still be expected in POS systems. As of this writing, at least one major computer manufacturer, Burroughs, is expected to announce new offerings in this area. Advances could also come from a little-known or unknown company; note the contributions made by UNI-TOTE, Information Machines, Ricca, Regitel, and Transaction Systems. The trend of development is unknown, though retailers now recognize what they need, and manufacturers now seem capable of supplying and are willing to supply retail information systems that are complete in themselves and that can relate closely to the salesperson.

8. DIFFERENT APPROACHES TO DATA COLLECTION

Although retail point-of-sale systems are the most active and representative area of data collection, they are certainly not the only application. The principles and techniques applied to retail data collection as well as many other techniques can be used to implement data collection in an unlimited number of types of systems.

INTEGRATED NETWORKS

Before describing the other types of systems, it is instructive to describe approaches to data collection which fail to meet the requirements of the definition given in the preface, and list what difficulties develop.

System One

Personnel who produce or transact the data (machine operators, sales clerk, etc.) write transaction data onto forms that are manually collected periodically and physically transported to a key-punch room where the data is transcribed onto punched cards and are later batch-processed by the computer.

DRAWBACKS

Extra (key-punch) personnel required.
Transcription process from handwritten form to punched card subject to errors.

Source-generated errors hard to trace and rectify considerable time lag.

Possibility of lost records.

System Two

Source personnel punch transaction data into cards via portable key-punches; cards are periodically collected and transported to computer area for batch processing.

DRAWBACKS

Manual key-punches slow and tricky to operate.
Source-generated errors still hard to trace and rectify.
Considerable possibility of lost documents or records.
Time lag.

System Three

Source personnel enter transaction data via Teletypes connected to time-sharing computer system. Computer updates data base after performing error checks on incoming data and notifies person responsible.

DRAWBACKS

Teletype not easy to operate, especially for large number of transaction types; not designed for source personnel.

Keeping the preceding imperfect approaches in mind, it is now possible to list the characteristics a system should have if it is to perform data collection quickly and efficiently.

BASIC CONFIGURATION

The basic system configuration should be a collection of remote data entry devices, or terminals, connected via wires or a communications link to a central device, or controller, capable of recording the data or transferring it to a computer. This requirement states, in effect, that communication rather than transportation should be the means of conveying information to a computer; communications provides the fastest, most direct path. The system can be made up of a large number of dissimilar devices as long as any two communicating devices are compatible.

Terminals

The data entry terminal should be designed with the main consideration being a smooth, efficient man/machine interface. The terminal should have a very simple data entry arrangement and preferably aid entry of specific, strictly formatted transaction types by appropriately labeling the keys and controls or by displaying appropriate messages. It need be designed for entry of only small amounts of data for each use. In short, the terminal should be designed for the quickest, most efficient entry of transaction data by source personnel, not people unassociated with the transaction.

Central Controller

The controller should not only record transaction data or forward it to a computer, but also perform continual checks on the validity of the incoming data. Checks should include those on transmission integrity, formatting, and completeness, and the failure of any check should be signaled back to the person sending the data, via his terminal. The data should not be recorded or transferred until all checks pass, or at least are flagged for later editing. Thus the person most suitable to correct an error is notified as it occurs, enabling immediate rectification. The controller should also add constant data to the transaction message, such as originating terminal number, date, and time of day. An additional benefit that could be provided by some of the controllers which transfer data to a computer is a means to query the data base at any time from a remote terminal.

APPLICATIONS

The primary benefit of systems meeting these requirements is that they provide management with the means for making decisions based on the actual status of the business. Once a computer has established an accurate data base and once it receives current data concerning all changes to that base, there is no limit to the types and quality of reports the computer can produce to communicate any facet of a business that a manager may require. The systems not only provide the means to maintain an accurate and current data base, they also simplify business procedures and personnel tasks by reduction or elimination of paper work, and by the simplification of many manual tasks.

The savings provided by reduced personnel costs and current, accurate data should, of course, be balanced against the cost of providing data collection equipment and perhaps the time required to enter data that would normally be ignored. However, the balance has weighed in the past, and is again beginning to weigh heavily, in favor of data collection systems, as is evident by the numbers of such systems becoming available and by the recent acceptance of such systems by certain classes of businesses, especially retail merchandising. Data collection system benefits are increasing while personnel costs are increasing and system costs are decreasing.

Chapter 9 describes industrial reporting systems, the other major application of data collection systems, while subsequent chapters describe other methods of system implementation, with special attention to new developments in data communications.

9. INDUSTRIAL SYSTEMS

Industrial data collection systems are designed primarily for the application of employee-attendance reporting, production status, inventory status, labor distribution, receiving and inspection, quality control, and machine maintenance. Terminals are designed for data entry by source personnel—the machinist for production status, the inspector for quality control—and normally contain a keyboard or set of dials, a badge reader, a card reader, and some sort of indication system. The keyboard is usually used to enter variable data such as a quantity or a machine number; the badge reader, to read an employee identification number; and the card reader, to read other prepared data such as a job identification number or a department number. The indicator system, or display, almost always notifies the operator whether the terminal is ready for use, when it is actually transmitting transaction data, and whether its message is correctly received by the central controller. Some terminals also provide an instruction display panel that is sequentially illuminated to guide data entry according to the application being transacted.

Messages from individual input stations are usually received by the central controller as single records, which are checked for accuracy, edited, and then transferred directly to a computer or recoded on a standard computer input medium such as punched paper tape or half-inch magnetic tape. The controller, or a device that interfaces the terminals and the controller, usually adds a source identity code, date/time data, and a message-validity code. A quick picture of what industrial data collection systems really are and how they are used can be given by describing three typical industrial applications using hypothetical equipment.

INDUSTRIAL APPLICATIONS

Attendance Reporting

Special terminals containing only a badge reader, some controls, and some indicators are usually used in this application, and function essentially as a time clock. Employees reporting to work push an IN button and insert an employee badge coded by punched holes into the badge reader. The terminal holds the badge, reads the coded employee number, and transmits the data to the central controller after being notified that the controller is ready. The controller checks the data for validity, formats it, and adds the sending-terminal identification number, the date, and the time of day. It then records the assembled data on magnetic tape and notifies the sending terminal that the recording is successful, causing the terminal to release the badge, light the appropriate indicator, and await the next transaction. When employees leave work, they operate the terminal in the same way, but push an OUT button. The resulting magnetic tape, which also contains data from other types of transactions, can be processed daily or weekly to report the actual number of hours worked by each employee.

Production Status

Full-fledged terminals, complete with badge reader, tab card reader, numeric keyboard, controls, and indicators are used in this application. Machinists starting work on a particular piece of work insert their employee badge, insert a tab card identifying the work piece, and use the numeric keyboard to enter a work area and a machine number. A similar procedure is used when the work is completed. The operation of the central controller is the same as that for attendance reporting.

If desired, quality inspection can also be made part of the production status records by having inspectors report results via the terminals. The inspectors operate the terminal in the same way, except that they use the numeric keyboard to enter the result of the inspection: whether the work is within tolerance, needs rework, or needs to be scrapped. The resulting magnetic tapes can be processed to prepare reports on the history and status of each work piece, the machinist, the time required, inspection results, and present location.

Tool Control

Tool-room clerks check out and check in each tool borrowed via a full-fledged terminal by inserting the borrowing employee's badge, inserting a tab card identifying the tool, and entering, via the numeric keyboard, a code identifying where the tool is being used. Tape processing can produce reports listing the location of each borrowed tool, the employee charged to it, when it was checked out, and when it was returned.

SYSTEM DESCRIPTIONS

Many different systems are available for industrial data collection, each varying somewhat from the hypothetical equipment used for the applications described above. Some have been available for several years, while others have been recently introduced or significantly redesigned to incorporate features made feasible by advances in circuit production techniques and reductions in cost. Representative industrial data collection systems are summarized in the subsequent paragraphs.

Control Data Transacter System

Control Data's transacter system is an industrial data collection system that accepts data supplied by badges, punched cards, and variable dials and entered at remote terminals. Data is collected and recorded on punched tape or magnetic tape at a central compiler (Fig. 9-1), which is

Fig. 9-1.   Control Data Transacter 1020 input station.

connected to the entry terminals via a 25- or 50-pair trunk line; this line may extend to 14,000 feet in length. A communications concentrator and converter can be used to connect input terminals to the compiler via tele-

Table 9-1. Control Data Transacter System Components

CODE	DEVICE	FUNCTION
1020	Input station	Reads one 15- and one 22-column punched card or badge and one 80-column punched card, one 10-position transaction core switch, six 10-position variable data switches, and wired plugboard data; sends data to compiler; synchronized clock
1021	Input station	Same as 1020 station but with nine 10-position variable data switches; no clock
0101	Badge reader	Reads one 15- or 22-column punched badge or card and wired plugboard data; sends data to compiler; synchronized clock
2010	Badge reader adapter	Interfaces trunk line carrying badge reader data to compiler
1091	Scale adapter	Interfaces weighing scale to trunk line; adds data to transaction message
2020	Paper tape compiler	Controls transacter system; collects transactions from input terminals; generates clock signals; punches collected data into paper tape
2025-1	Magnetic-tape compiler	Same as 2020 but writes on magnetic tape; also includes input/output typewriter
2025-2	Magnetic-tape compiler	Same as 2025-1 but adds magnetic-tape transport and 150 line/min printer
2040	Compiler shunt	Automatically transfers input of one compiler to second compiler upon detected malfunction
2042	Automatic split-load feature	Enables 2040 shunt to automatically share compilers upon simultaneous trunk-line demands; reverts to single compiler operation after settable time delay
2070	Concentrator	Interfaces input terminal trunk line to Bell Systems 402C data set
2075	Converter	Interfaces Bell Systems 402D data set to compiler
2015B	Cable repeater	Reshapes trunk-line signals to extend maximum terminal/compiler distance to 28,000 feet

phone lines and data sets, without restriction on transmission distance. Any number of input terminals can be connected to a compiler, but a maximum of 36 is recommended to keep system response time to a few seconds under heavy loading conditions. A system can be configured with two compilers, allowing parallel operation under heavy loading and backup operation when necessary; switching can be accomplished automatically with the shunt option.

The compiler adds up to 17 characters of compiler clock and fixed data to the data sent from the input terminals. Clock data comprises such information as shift, day, hour, and minute, while fixed data, generated by

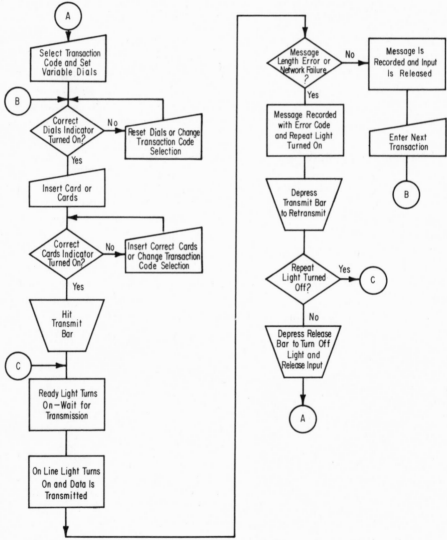

Fig. 9-2.   Operating procedure for the Transacter data collection system.

a wired plugboard, comprises program instruction codes, location codes, and other information to be used during subsequent tape processing.

Common system components and their basic functions are listed in Table 9-1. This system was introduced in 1959; since that time it has undergone several revisions to keep current with customer requirements and new electronics technologies. Since magnetic tape is now used much more widely than paper tape for computer input, the paper tape compiler has been dropped from production. However, the concentrator and converter were introduced after the original system release, to provide the capability to transmit data from a group of terminals over a voice-grade phone line rather than a multiwire trunk line, thus satisfying a demand that is rapidly increasing. More about this data communications aspect will be presented in Chapter 12.

Terminal operation is diagrammed in Figure 9-2. After selection of a specific application on the leftmost terminal dial, the compiler applies application-dependent validity checks on dial and card data, lighting the appropriate indicators for any detected errors. Thus, even though the terminal gives no direct indication of what type of data should be entered for a specific application, it enforces correct application by not operating until all validity checks have been passed. The compiler also checks for correct message lengths and performs a read-after-write check on each magnetic-tape record.

IBM SYSTEMS

IBM 357 System

IBM has introduced several versions of data collection systems. The IBM 357 data collection system is an early, simple system that is still available. It is designed as an off-line system. Input (Fig. 9-3) can be read

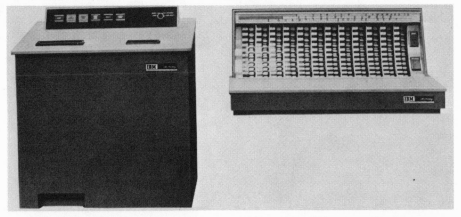

Fig. 9-3.   The IBM 357 input station (left) and 12-slide manual entry unit.

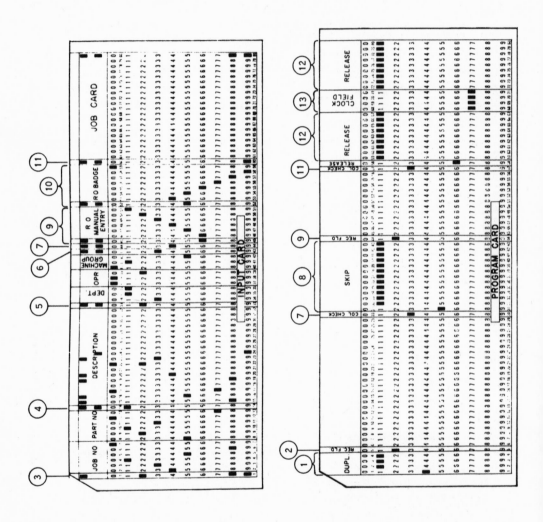

Fig. 9-4. Typical input, program, and output cards for the IBM 357 data collection system.

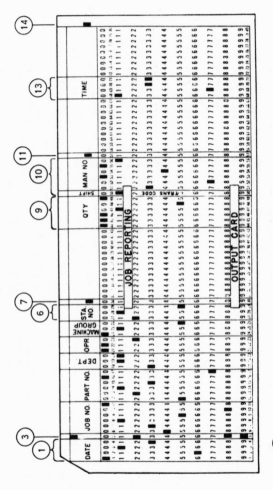

Date is automatically duplicated through all the cards.

Initiates reading and punching from the input station.

Transaction Code.

Start and stop skipping of input card field.

Input card code causes station number to be punched.

In-sequence check column.

Causes skipping in the output card.

Input card punches and program card punch cause slide settings 6-1 to be punched.

Input card punches cause badge columns 4-9 to be punched.

Input card punch causes in-sequence check and a correct or incorrect punch in the output card.

Program card punches cause output card columns to be skipped.

Program card punches cause the time to be punched.

Correct or incorrect code for all in-sequence checks.

69

from standard 80-column prepunched cards, from punched identification badges, or from up to 12 manually set 11-position slides. Designed card columns, badge columns, and/or slides can be read or not read, depending on column-dependent "instruction codes" that are contained in a central input control unit.

The off-line output medium in this system is the punched card. A modified version of either the ibm 24 or 26 card punch is used to produce output at a rate of up to 20 characters per second. The input control unit enables up to 20 input stations to be cable-connected to one card punch. Switching is accomplished by sequential scanning of "ready" input lines; from 35 to 700 msec are required to search through all 20 input stations for a waiting station. The output format can be controlled by a combination of program card and plugboard wiring, and data can be punched into all or any of the 80 columns in each output card.

Transmission of data is automatic upon insertion of a card or badge into the proper reader slot. Variable data can be set up off line and then transmitted under control of the card reader. Discrepancies in transmitting (either by commission or omission) are indicated by lights and up to 15 seconds are allotted for making any necessary corrections to the current transaction; otherwise, the entire transmission must be repeated. A transaction can consist of any number of cards in sequence, so the 15-second wait is valuable in case an error is made near the end of a long transaction.

The 357 system has several features that can be incorporated into an installation at the user's request. Among these are the following options:

1. Portable manual entry units (data cartridges), which permit off-line composition of variable data
2. Readout clock for automatic recording of time at the end of each transaction
3. Switch control for automatically switching to a backup punch

The data cartridges enable a user to preset up to 12 digits of numeric data away from the input station. The preset cartridge is inserted into a 357 input station during transaction data entry. The use of the cartridge reduces waiting time and data entry errors.

The 357 is extremely simple to operate for standard transactions, and its simple design results in terminal costs on the order of $1000 to $1750 and central control unit costs of about $3000 to $7000, costs that are significantly lower than those of most card-reading terminal systems. Much of the system simplicity is due to the plugboard and program card method of control, a low-cost method but one for which system modification is tricky. Typical input, card punch program, and output cards for a job-reporting application are shown in Figure 9-4.

IBM 1030 System

The IBM 1030 data collection system, introduced in 1964, is a more advanced system than the 357, offering significant differences in most system aspects except that of data entry.

The system is designed primarily for two-way communication between remote plant locations and a central processing area. Input can be from prepunched cards, identification badges, manual entry slides, or preset data cartridges. A variety of input units are available for handling various combinations of these input media.

The input stations are available with two types of line capabilities. A "control" station (model A) operates over two-wire communication lines, while a "satellite" station (model B) transmits over a multiwire cable attached to a control station. This flexibility permits a wide variety of

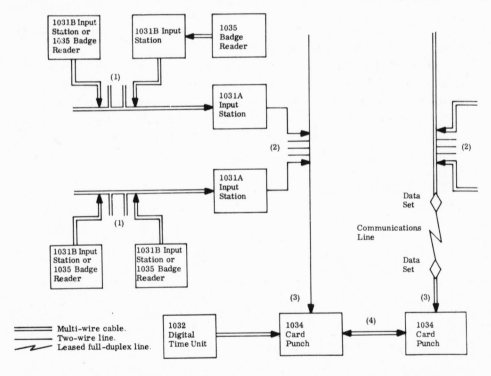

(1) A total of eight 1031B's and 1035's (no more than four 1035's) can be connected to a 1031A. See Paragraph 2 for limitations on the number of columns of badge data that can be read.

(2) A maximum of ten 1031A's can be connected to a 2-wire communications line.

(3) A maximum of 24 1031A's, 1031B's, and 1035's, in any mix, can be connected to a 1034.

(4) Stand-by interconnection.

Fig. 9-5. Configuration possibilities for off-line 1030 data collection systems.

system configurations with varying combinations of control and satellite stations. The maximum number of control and satellite stations per communications line is

1. Ten control stations per communication line.
2. Eight satellite stations per control station.
3. Four badge readers per control or satellite station.
4. Total of 24 input units (control and satellite stations and badge readers) per two-wire communications line.

Figure 9-5 shows configuration possibilities for the 1030 system. Note that control stations can interface to the central processing area via voice-grade telephone lines. The 1030 system transmits all input data to the central processing area at 60 characters per second.

For off-line applications, punched card output can be produced by connecting one or more 1034 card punch units to the transmission line(s). One punch can serve as output for any combination of up to 24 input units. Input unit polling, as well as parity and message-length checks, are performed at the card punch.

The output message format is shown in Table 9-2. One message is normally punched into each output card. Transaction codes can be wired for selective reading of the input. An optional "packed card" feature provides for punching multiple messages into one card during specified time periods (usually for attendance recording). An end-of-message code is automatically punched after the last input called for by the transaction code has been read.

The 1034 card punch, in normal operation, punches transmitted data at 60 characters per second from up to 24 input units connected to a single

Table 9-2.   IBM 1030 Output Message Format

| | NUMBER OF CHARACTERS | |
FIELD	ALPHANUMERIC	NUMERIC ONLY
Date and/or network ID*		1–8
Input station ID*	1	
Transaction code		1
Card data †	1–78	
Badge data †		1–10
Manual data †		1–12
Time ‡		4
Error code *	1	

*Added by 1034 card punch.
†If entered.
‡Added by the 1032 digital time unit.

2-wire line. For standby operation, up to 48 input units connected to 2 lines can be served by a 1034 card punch if additional addresses have been provided in the polling sequence. Time data, and fixed data entered by the 1034, are punched at 80 columns per second.

The 1030 system can also be configured into an on-line system with a transmission control unit at the computer site to control message reception, character assembly, polling of communication lines, and other related functions. These units can handle from 4 to 20 lines. Some models allow connection to an IBM Series 1400 computer, and others allow connection to IBM/360 computer systems. Optional automatic time equipment can record the time of day for each transaction.

Each control station provides for the connection, by an additional multiwire cable, of up to nine printers. The printers, in conjunction with an input control station, give the 1030 system on-line inquiry capabilities

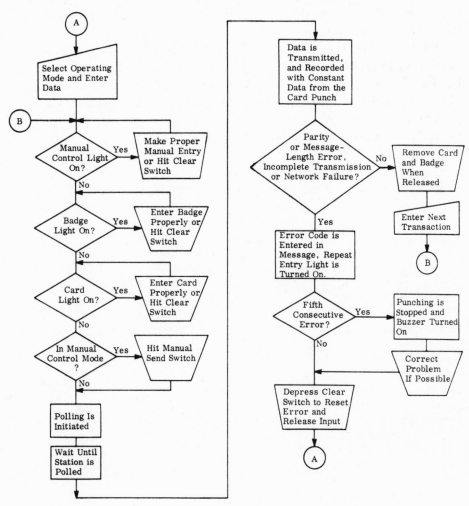

Fig. 9-6.   Operating procedure for the 1030 data collection system.

to and from a computer. Up to nine printers can be connected to an input control station, but no more than 24 can be connected across any one two-wire line.

Despite the increased capabilities provided by the 1030 over the 357, the 1030 is still very simple to operate. The operating procedure is shown in Figure 9-6; for badge and card transactions, all the operator need do is enter the badge and card and wait until they are released. However, 1030

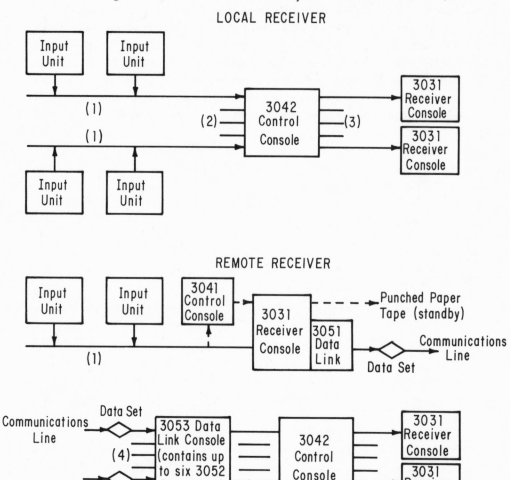

(1) Up to 20 input units (in any mix) per transmission cable.
(2) Up to 20 transmission cables per Control Console.
(3) Up to 22 Receiver Consoles (2 standby units).
(4) Up to 6 transmission cables.

Fig. 9-7.   Friden COLLECTADATA configuration possibilities.

component prices are significantly higher than those of the 357 system and are also higher than those of many competitive systems.

FRIDEN COLLECTADATA

The Friden COLLECTADATA is a data collection system that reads alphabetic and numeric data from punched cards and numeric data from identification badges and dial settings.

Input units for the COLLECTADATA system (Fig. 9-7) are the badge readers, which read identification badges only, and transmitters, which read a punched card and a badge or two punched cards. Data is transmitted and recorded at 30 characters per second.

A COLLECTADATA system can comprise up too 400 badge reader and card transmitter stations in any mix connected to as many as 20 transmission cables. The cables terminate in receiver consoles to generate punched paper tape or provide data directly to an on-line computer: for example, UNIVAC 1108, GE 225, Honeywell Series 200 computers, and IBM/360 computers. When the receiver console or computer system is located more than 2 miles from the transmitters, Friden data-link units and data-link consoles are required for transmission over a leased voice-grade communications line.

Some typical system configurations are shown in Figure 9-8. For attendance recording, data from a badge is normally read into storage immediately and transferred to the receiver console. This allows the operator to step away from the reader immediately. Data from the next badge cannot be read until storage is cleared by transferring the data to the receiver console. If there is a delay, the next badge will be held in the reader until its data can be read into storage. The reading of the badge is controlled by plugboard wiring or by a code punched in the card that is read as part of the data collection operation.

Prepunched tab cards containing data fields and control codes are used for entry of prepared data. Each punched card contains a card code (a 12, 0, 1, or 2 punch to permit checking for proper insertion), a code that initiates the reading of dial settings, a badge readout code (if required), descriptive information (e.g., item name and number), and an end-of-data code. Up to 76 characters of data can be punched into a single input card, and an additional 78 data characters can be punched into a second input card for those transmitters equipped with two card readers.

The transmitters are equipped with a seven-sided transaction selector bar that can be rotated to select one of seven types of transactions to be recorded and to provide instructions for entering the transaction. The name of the transaction, a yes-or-no indication for badge entry, the name of the

ATTENDANCE RECORDING MODE   DATA COLLECTION MODE

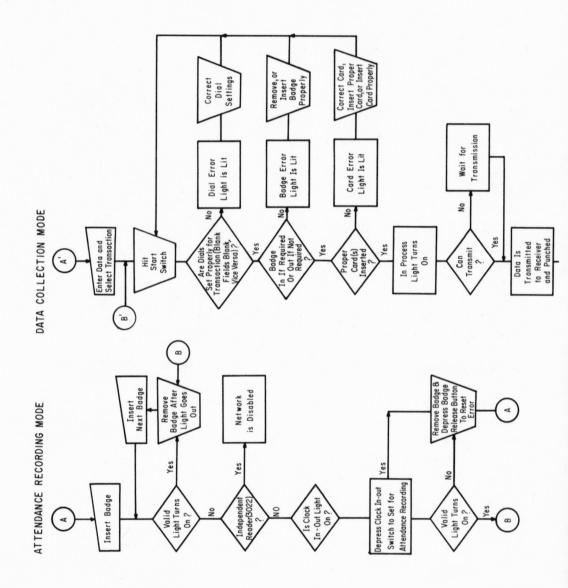

76

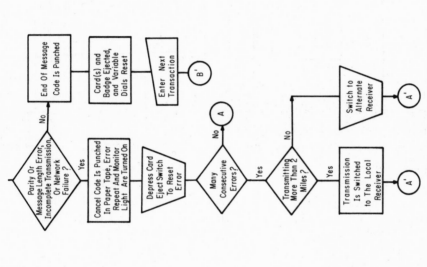

Fig. 9-8. Operating procedure for the Friden COLLECTADATA system.

77

punched card to be entered, and the name of the field for each dial setting are shown in 13 display windows.

Ten dials can be set by the operator to enter variable numeric information such as the number of pieces processed and the machine number. Eight more dials, located under a lockable cover, can be set to enter fixed data such as station number and department number. These dials, like the variable entry dials, can be set for the digits 0-9, a blank, or one of two special symbols.

Time and date messages are generated by the central time transmitter contained in the control consoles. Each receiver console records data received from up to 20 transmitters and badge readers at 30 characters per second and controls the transmission sequence of the input units. Whenever one input unit is transmitting, all other connected units are prevented from transmitting. After an input unit finishes a transmission, units farther from the receiver are monitored. If more than one has requested transmission, the earliest request is serviced. The monitoring system uses electric impulses rather than stepping switches. The maximum wait time for an input unit in seconds is one-thirtieth (30 characters) the sum of the maximum number of characters per message that can be transmitted by all other units connected to the receiver.

To bring its COLLECTADATA system up to date, Friden has recently announced a new system configuration incorporating the Model 20 central processing unit of its new System Ten computer. The Model 20 processor simply replaces the control console and receiver console units of the original system. Thus, terminals are connected directly to input/output channels of the processor through a special terminator that converts the 90-volt terminal signals to 5-volt signals used by System 10.

The Model 20 processor controls the input terminals; controls the flow of data from the terminals by an elaborate, balanced, traffic-dependent cable-sharing system; performs error checks; adds fixed and time data; and records the formatted data to computer-compatible magnetic tape. System malfunctions or operating errors are reported immediately via a console typewriter at the processor.

The use of a single processor to replace several system components results in significant system simplification, and it appears that Friden will gradually replace existing COLLECTADATA systems with COLLECTADATA systems incorporating the System Ten processor.

COLORADO INSTRUMENTS SYSTEM

Colorado Instruments' data collection systems (Fig. 9-9 A, B, C) enter source data via one of various C-DEK data input stations or badge reading

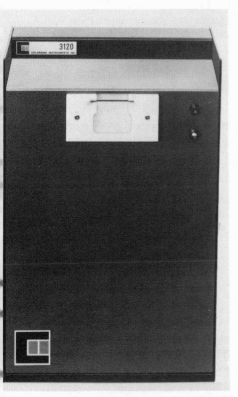

Fig. 9-9.　Colorado Instruments system components.

stations, and then transmit it, at up to 1000 characters per second, to a centrally located or remote paper tape recorder, card punch, magnetic-tape recorder, or computer system. Input may be alphanumeric data pre-punched in 80-column tab cards or 22-column identification badges. Variable alphanumeric data is entered via lighted key columns. Semiconstant numeric data such as department number can be entered, using dials. Constant data such as data input station identification number is automatically entered by the input station.

The system comprises data input stations, in any combination, and a central controller with a capacity of up to 120 stations. The input stations accept and transmit data from badges, cards, variable key columns, and dials. The central controller performs sequential polling of the input stations and all error-checking functions, and prepares the incoming data for interface to the output media. The multiplexer adds any required prefix and suffix data and date/time information to the message from the input stations. Figure 9-10 shows configuration possibilities for one system.

The stations may be connected to the central processor either via a multiwire cable system or two-wire pairs. The multiwire cable systems provide a higher data transmission speed and reduce terminal and controller cost, requiring less expensive communications circuitry, but the twisted-pair systems result in low installation and calling costs, and provide unrestricted component mobility.

The design of the c-DEK input stations is intended to make data entry procedures as close as possible to entry of data on paper forms, a procedure with which a typical operator is most familiar. The variable-data keyboard is an array of 10-key numeric key columns, one column for each digit of variable data. For each group of applications, a custom-designed overlay mat, or template, is provided to assist the operator in making correct entries by identifying and grouping data to be entered in the various key columns. It also identifies individual keys in a separate transaction-type column.

The standard variable-data keyboard comprises 13 columns, which can be extended to 26 columns by using standard modules. A two-position HOLD/RESET switch is available for insertion in any variable-entry key column position where transmission of certain data fields may occur repeatedly. The HOLD/RESET switch, which occupies the 12th key position, may be in either the HOLD position to retain the data field for subsequent transmissions, or in the RESET position, which clears the data field after each transmission. Semiconstant data such as department number and date can be entered, using up to eight thumbwheel switches.

The input stations include as standard equipment an 80/22 column card reader, which reads up to 78 columns of an 80-column card and up to 20 columns of a 22-column identification badge. Card columns 1 and 80

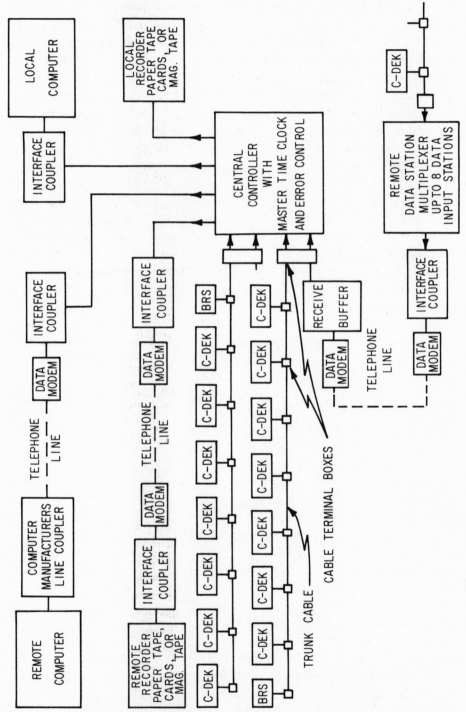

Fig. 9-10. Colorado Instruments data collection system configuration possibilities.

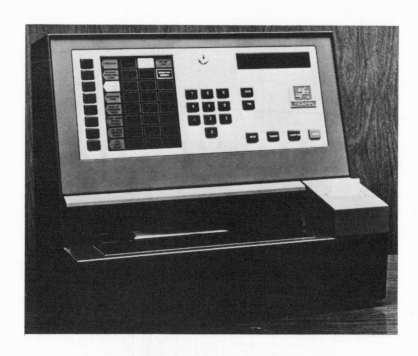

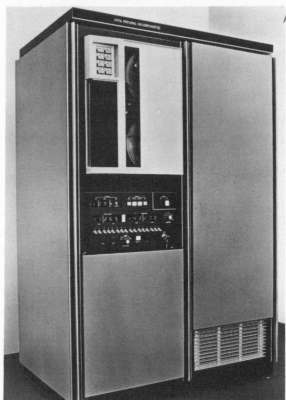

Fig. 9-11.   The DPI source data management system components (partial).

83

and badge columns 1 and 22 are used for control purposes. Any transaction can control the type of badge and card that is entered in and accepted by the c-DEK. For example, a transaction used for "inspection reporting" might require that only an inspector's badge will be accepted for the subsequent transmission. This is accomplished by comparing preassigned codes with the code punched in column 1 of the badge; if they match, transmission is completed. If they do not match, a red indicator lamp below the badge reader section indicates that an improper badge has been entered and transmission cannot be completed. Card column 1 is also used to control the type of card entered for any specific transaction. Card and badge data may be numeric or alphanumeric.

The entry procedure is fast and straightforward, and reduces the possibility of keying errors. Extensive error checks are perfomed to ensure accurate data entry and transmission. Ten different transactions and their corresponding entry format are available. For each transaction, a preprogrammed logic card causes various red column-indicator lamps to light and show the operator where data *must* be entered. As data is entered, the lamps are extinguished, indicating proper entry. The indicator lamps also show where data entered is not needed. Transmission cannot be accomplished until the proper data, and *only* the proper data, has been entered correctly and all indicator lamps have been extinguished.

MOHAWK DATA SCIENCES 4400 SYSTEM

Mohawk Data Sciences sells Colorado Instruments' data collection system under the name of the MDS 4400 source data gathering system. All components are the same except that Mohawk modifies the central controller and adds a Mohawk 1100 or 6400 data recorder to produce the data collection magnetic tape. Otherwise, the characteristics of both systems are essentially the same.

Early in 1971, Mohawk acquired Colorado Instruments, but for the present they will continue to market Colorado Instruments' original system under the name of Colorado Instruments; Mohawk's modified system will carry their own name.

SIERRA RESEARCH SDA-770

The Sierra Research SDA-770 incorporates a custom-programmed Honeywell H-316 minicomputer in the central controller for direct control

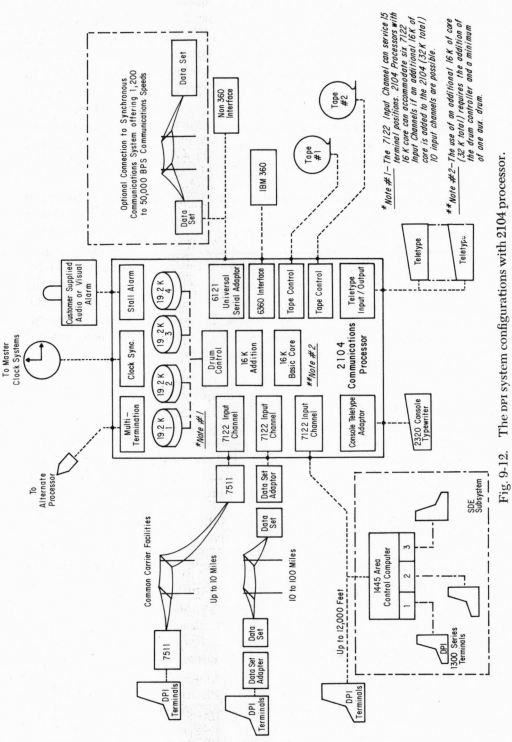

Fig. 9-12.　The DPI system configurations with 2104 processor.

*Note #1.– The 7122 Input Channel can service 15 terminal positions. 2104 Processors with 16 K core can accommodate six 7122 Input Channels if an additional 16K of core is added to the 2104 (32K total) 10 input channels are possible.

**Note #2.– The use of an additional 16K of core (32 K total) requires the addition of the drum controller and a minimum of one aux. drum.

85

of terminal and system operation. Up to 128 terminals can be interfaced to the central controller, with up to 15 terminals utilizing a single twisted pair or telephone line. Terminals, which are custom designed for their particular applications, accept badge, punched card, or keyboard input of data and use a programmed instruction display panel to guide the terminal operator. Keyboard-entered data is retransmitted to the sending terminal and displayed after it is received by the central controller for operator verification before final entry.

An internal column printer and/or external page printer at each terminal provides a hard-copy record of data transactions; terminals also display answers to operator inquiries. Worst-case system-response time for any terminal is 200 msec. The central controller, which performs extensive error checking on all data received, also collects, formats, and assembles complete transaction data before recording it on the output unit. Besides adding time, date, and remote terminal address to the output record, the central controller optionally performs data reduction and partial processing of the collected data, with special software and an optional disk or drum storage unit. Transaction data can be output to any of the following output units: magnetic tape transport, tab card punch, paper tape punch, data set, and computer communications interface. An ASR 33 Teletype is used to provide control access to the central controller.

The SDA-770 is a relatively new system, released late in 1970, and several of its features are not found on systems described earlier in this chapter. One of the most important features is the operator instruction display, which, once a transaction type is selected, sequentially illuminates instruction messages under rental controller software control to guide and enforce proper operating sequence. Variable-data entry is accomplished with a 10-key keyboard arranged like that of a Touch-Tone telephone. Data entered by the operator is not displayed by the six-digit numeric readout; instead, the display shows data as received by the controller, enabling the operator to verify directly that the data as it will be recorded is correct. The printers can be used to provide audit trails before computer processing of the transaction data, and the capability to respond to operator inquiries via the numeric display provides access to the collected data base, which is normally inaccessible prior to processing on conventional systems.

The entire SDA-770 system is modular and can be configured, within limits, to a large number of specific applications. For example, the terminals can be equipped to accept and transmit variable data for external hardware devices such as A/D converters, scales, and alarms.

DATA PATHING DPI SYSTEM

Another relatively new system developed by Data Pathing Incorporated (DPI) provides instruction display and inquiry response capabilities like those provided by Sierra Research's SDA-770. This system has been recently updated and expanded to include a comprehensive line of input terminals and a convenient approach to user software. It is described in some detail, as it is representative of the comprehensive approach to data collection that will be provided by some of the new systems yet to be introduced.

The DPI source data management system formats, edit checks, and records on magnetic-tape data entered at remote terminals via keyboards and/or prepunched badges and cards. The system can also route the data recorded on magnetic tape to an on-line computer or a communications interface. Several types of remote terminals are available, including one designed exclusively for badge entry for attendance reporting, another with a card reader and numeric keyboard added to a badge reader designed for entry of a number of transaction types, and a group designed for alphanumeric keyboard data entry. Data entered at remote terminals is collected at a central communications processor, which may vary in input terminal capacity from 30 to 150 terminals. System configuration for a typical DPI system is depicted in Figure 9-12. Characteristics of the various terminals and communications processors are listed in Table 9-3.

The system is controlled by the communications processor, operating under control of customized software provided with the system, called the data collection operating system (DCOS). Functions performed by the software include code conversion, numeric data checking and editing, adding constant data, output data formatting, error checking, tape unit control, and on-line computer interface control. The DCOS also maintains application programs that control terminal and processor operation during data entry for specific transaction types.

Normally, the application program for attendance reporting is resident at each terminal unless an alternate program is selected by depressing the appropriate program button on the terminal. This action causes the application program for the selected transaction type to be stored into the terminal memory for the communications processor. The application program displays instruction messages at the terminal in a preprogrammed sequence to guide the operator in proper transaction entry. Messages call for in-

Table 9-3.   DPI Terminal Characteristics

TERMINALS	INPUT MEDIA	DISPLAY
1101 Badge reader	10-digit punched badge	Ready, off-line, transmitting, repeat indicators
1203 Multi-function terminal	22- and 80-column punched cards; 10-key numeric keyboard	Instruction display; 5-digit numeric data or time display; enter, repeat, transmit indicators
1204 Multi-function terminal	10-digit punched badge; 22- and 80-column punched cards; 10-key numeric keyboard	Same as 1203 terminal
1301 SDE terminal	13-key numeric/alphanumeric keyboard; codified input capability	Numeric data display matrix (alphanumeric optional); instruction display
1302 SDE terminal	10-digit punched badge; 13-key numeric/alphanumeric keyboard; codified input capability	Time display; instruction display
1304A SDE terminal	10-digit punched badge; 13-key numeric/alphanumeric keyboard	16-character alphanumeric data display; instruction display
1304B SDE terminal	10-digit punched badge; 54-key full alphanumeric keyboard; codified input capability	Same as 1304A terminal
1311 Key processing terminal	029 alphanumeric keyboard; 16 additional function keys	Single-character alphanumeric data display; 18-position operator instruction panel
1314 Key processing terminal	Same as 1311 terminal	16-character alphanumeric data display; 18-position operator instruction panel
3101 CRT terminal	Keyboard: 42 alphanumeric characters, 64 ASCII graphics symbols, 12 edit keys, 4 special symbols	20 lines of 50 characters each; 18-position operator instruction panel
3102 CRT terminal		24 lines of 80 characters each; 16-position operator instruction panel
2740 Inquiry/ retrieval terminal	IBM-compatible Selectric keyboard	Printed page
5101 Batch card reader	80-column punched cards	
5100 Line printer		80-, 132-character printed lines
5210 Printing card punch		80-column punched cards

sertion of a badge or specified punched card or entry via the keyboard. They also identify the data type by calling for item quantity, status code, and operation number, for example. Most DPI terminals permit direct selection of up to seven different application programs, with others available by coded request via the numeric keyboard. The backlighted instruction masks are removable to facilitate changing the set of programs available to a terminal. Thirty-one different masks can be handled by the communications processor, with a maximum selection of 115 programs.

Application programs also provide a series of edit checks on input data. Predefined editing routines check for equality or nonequality of input data to a known constant or of two separate data inputs to each other, check that a numeric value is within defined upper and lower limits (four digits maximum), or check that the input data is a specific type (alphabetic, numeric, alphanumeric, or blank). Badge and keyboard data is also checked for validity of data codes.

The results of edit checks can be used to determine the actual steps required to enter a transaction. Software can "chain" two or more programs together, selected by edit checks, to form a single application program. Software also checks that all elements required for a transaction are entered. After all checks have been performed successfully, the communications processor so notifies the terminal by returning it to the attendance-reporting mode and releasing any inserted badge. Besides performing data checks, the processor adds date, time, sequence, and other data to the original transaction data and formats it according to DCOS preprogrammed specifications before it is routed to the system output device.

Data Pathing, Inc., supplies a software program generator called AIDE, which enables the customer to change DCOS application programs and to generate new software systems as desired by a simple card-to-tape routine.

Besides providing data collection functions, the DPI system may be equipped to control tape-to-tape data transfer over voice-grade communications lines from a communications processor to another processor or other communications equipment. The processor may also be equipped to handle two-way terminal/processor communications, which enables inquiry feedback to terminals, processor-controlled data input to teletypes, and on-line transaction validation. In-plant message switching, also available, permits Teletype messages to be input to the DPI processor and subsequently routed to another Teletype, the system output tape, or the on-line processor.

Terminals for remote data entry comprise a group intended for reporting of attendance and production data, a low-cost group intended for more multipurpose individual or semi-individual data collection, and others intended for general keyboard to magnetic-tape data entry or inquiry/response functions.

The first group of remote terminals, comprising the 1101, 1203, and 1204 terminals, is connected to the communications processor through an input channel device located in the processor. Terminals are connected via two-wire cables, with as many as 15 terminals per input channel and as much as 12,000-foot total cable length for the set of terminals, including trunk and spur lines.

Up to 10 input channels for a total of 150 terminals can be connected to the processor with the 16K additional core-memory option. Communications distances may be extended to 10 or 100 miles by using a line conditioner or data set adapter. The limitation per input channel remains at 15 terminals for these options.

Input to this group of terminals is via badge, 22-column card, 80-column card, and/or numeric keyboard, depending on the type of terminal and transaction. Code types and digit capacities of the input media are listed in Table 9-3. Required media are requested by illuminated messages on the display panel. Badges are locked in place, once inserted into the terminal, until the communications processor acknowledges to the terminal that the transaction has been successfully recorded; badges may be inserted either forwards or backwards, with the processor automatically performing data reversal if necessary. The badge-in-scanner option permits badge interpretation (inserted in any of four orientations) by the 22-column card reader on the terminal. Data keyed into the numeric keyboard is displayed by a five-digit numeric display panel as it is entered. Up to five digits may be entered for each program segment, with a limit of 10 digits per application program or 80 digits for a chained group of programs. All positions of 22- and 80-column cards are read by the terminal card scanners. Multiple cards can be read for single transactions by chaining application programs.

The second group of remote terminals consists of the Series 1300 terminals. Instead of connecting directly to the input channel of a communications processor, they communicate directly with an area control computer, which in turn is connected to a communications processor input channel. The area control computer provides terminal control functions, storage for lamp-lighting applications programs, and data buffering and editing. Completed messages are routed to the communications processor after they have been edited. Up to 45 Series 1300 terminals may be connected to one area control computer, with up to 15 sharing one to five 9-conductor trunk lines of 1000 feet maximum length. An area control computer interfaces to a communications processor in the same manner as a Series 1200 terminal, making it possible to connect any combination of area control computers and Series 1200 terminals, up to a total of 15, to one communications processor input channel.

Data entry via the Series 1300 terminals is primarily from a simplified 17-key numeric or full 54-key alphanumeric keyboard. The simplified keyboard permits entry of alphabetic and special characters by the use of three shift keys to change the functions of the normally numeric keys. The full keyboard enters alphabetic characters directly. Badge data input is also accepted by some of the terminals. Certain data such as transaction type and employee number can be entered by a codified input capability available on some terminals, which permits depression of a single numeric key to represent items that have been precoded as indicated on the terminal's front panel. Data entered from the keyboard is limited to 16 characters per field and 112 characters per transaction. A 16-character matrix display on most terminals displays numeric characters as they are entered; optionally, the display may be also equipped to represent alphabetic and special characters.

When inquiry/response functions are incorporated into the system, the terminal display functions as a readout device. Editing is performed on input data in the same manner as for the Series 1200 terminals, but by the area control computer rather than the communications processor. Records entered by the terminals are stored on a drum in the area control computer until they are forwarded upon request of the processor.

Data may also be entered into the Data Pathing system from a master clock, a console typewriter, a CRT terminal, a Teletype, or a magnetic-tape unit. Terminal operation is simplified by the instruction lamp-lighting program sent to a terminal after selection of a particular application. Instructions tell the operator what to do during each step, and edit checking ensures that the correct type and quantities of data are entered. For variable-data input via the numeric keyboard, the ENTER lamp is illuminated once the required number of digits for a particular field, as predefined by the application program, is entered. The operator can then verify that the data entered, as shown by the numeric display, is correct by depressing the ENTER button, or he may reenter the data after depressing the CLEAR button. When all data required for a complete transaction has been entered, the TRANSMIT button is illuminated; by depressing that button, the operator initiates actual data transfer to the processor. If the processor verifies that the transaction data is complete, is of the correct type, satisfies the edit checks, and is correctly recorded, it acknowledges the successful transaction by releasing the badge at the entering terminal and returning the terminal to the attendance mode.

The data collection is started, maintained, and debugged from a control panel located on the communications processor in conjunction with the console Teletype.

The Data Pathing system represents an advancement in the development of data collection technology by providing a well-designed man/ computer interface. The terminal data and instruction displays are very closely coordinated with operator functions during data entry, and the data display may also be used to obtain requested information from the collected data base.

The variety of system options permit a user to configure the system to his own requirements. Besides providing a large number of input terminals, the system allows selection of a number of different configurations of input communications facilities, central processor designs, and output devices.

Incidentally, Data Pathing's 2104 central communications processor is also used as the central controller for Transaction Systems' point-of-sale registers, described in Chapter 5.

10. OTHER APPLICATIONS

BURROUGHS SYSTEM

The data collection systems described thus far are not limited solely to industrial and retailing applications. As an example, Burroughs manufactures data collection terminals (which will be described in Chapter 13 because they are not supplied as part of a complete system) that provide similar capabilities to some of the terminals already discussed: they have a simple variable-data keyboard, read punched cards and badges, and display computer responses to inquiries (via a printer). Burroughs' literature lists the following applications for its terminals:

1. *Manufacturing:* storeroom receipts and disbursements, receiving/inspection, shop floor order control, purchasing, machine maintenance, quality control, metals laboratory reporting, general and cost accounting, sales offices (order entry), consigned stock inventory.

2. *Services:* auto dealer accounting (parts inventory, customer billing, auto leasing inventory and accounting), professional services billing (doctors, dentists, attorneys), hotel reservations systems, utility billing, grain elevator co-op, (accounting and inventory).

3. *Wholesale distribution:* receiving, sales-order entry, purchasing, stock-picking verification.

4. *State and local government:* tax payment receipting, inventory control, capital asset accounting, transportation pool assignment, payroll inquiry.

5. *Education:* library, stores inventory, receiving, purchasing, grade reporting, student record inquiry, audio-visual inquiry and reservation.

6. *Financial:* credit inquiry, account balance inquiry, holds, stop payment.

7. *Medical:* nursing stations, admitting office, pharmacy inventory, clinical lab reporting, general accounting, cash receipts, nursing home patient billing and general accounting.

As a specific alternate application, Colorado Instruments distributes its fundamentally industrial C-DEK system in a specially designed, integrated system for use in libraries by librarians. The system, called circulation input recording center (CIRC), is modular in design so that it can be adapted to meet specific requirements of any library.

The system may be configured as an on-line or off-line system, the off-line system recording on a central recorder and the on-line system transmitting to and receiving data from an IBM/360 computer with automatic backup recording on a magnetic-tape unit. The on-line system includes printers located near each input terminal, which receive data from the computer. Typically, one or more input terminals are located at each circulation center. The terminal includes an 11-column numeric keyboard, a 10-key column of transaction buttons, controls, an 80-column prepunched book card reader, and a 22-column prepunched patron-identification badge reader. The functions normally performed by the terminal include book charge and discharge, entering and removing materials from the reserve list, creating a hold list, monetary data input for fines, and payroll attendance recording.

COLORADO INSTRUMENTS SYSTEM

Colorado Instruments has defined the following procedures for circulation charging and discharging.

Charging Books

Before going to the stacks the patron consults a current circulation list to see if the desired book is in circulation. After locating the desired book, he takes it to the circulation desk and presents the book and his identification card. The circulation clerk selects the CHARGE OUT transaction, keys in a specific loan period, and inserts both the book card and the patron's identification card into the circulation center. The combined data is read and recorded at a central magnetic-tape recorder or transmitted on-line to the computer. Exact time and date is automatically added to each charge-out message. A CIRC 80/22 option automatically prints a date-due slip containing call number and date due. Another less expensive method of preparing date-due slips is also available. This method requires the circulation clerk to stamp required number of date-due slips each morning. When a book is charged out, the clerk enters the date-due slip in the book pocket. If a library guard is used, he can mark the date-due slip as each book is checked through. This eliminates the possibility of a slip's being used for more than one book.

A special C-DEK feature can be provided for holding the patron's badge in position for charging multiple books. Another feature eliminates the need to key-in a standard loan period for each charge-out. Still another feature allows multiple charge and discharge transactions to be performed without having to key-in the transaction for each message.

In an on-line system, the computer immediately checks its records to see if this transaction is valid. Items such as an invalid identification card, no current patron address, too many books charged, or fine due may be checked and any irregularities immediately printed out at the circulation desk for corrective action. All data is then used to update the circulation list.

In an off-line system, all pertinent data is recorded on magnetic tape for subsequent computer processing.

In both off-line and on-line systems, all circulation data is accumulated for later statistical analysis.

Discharging Books

To discharge a returned book, the circulation clerk selects the DIS-CHARGE transaction and inserts the book card in the circulation center. The combined data are read and transmitted to the central recorder or on-line to the computer. Again, exact time and date of the transaction is automatically added to the discharge message.

In an on-line system, the computer immediately searches its records and clears the corresponding BOOK CHARGE transactions. If the particular book has been placed on the recall, hold, or reserve list, a message is printed out at the circulation desk to notify the clerk that the book requires special handling.

In an off-line system, the data is recorded for subsequent batch processing. In either instance, the computer removes the record from the next day's circulation listing. Overdue notices are automatically calculated by the computer. These notices are printed on postcard stock, ready for mailing to the patron. If the book is overdue when returned, the computer automatically creates a fine-due notice on postcard stock, ready for mailing to the patron.

System Capabilities

Other procedures are followed for reserves, special charge-outs, holds and recalls, fine payments, inventory inquiries, fine status inquiries, attendance reporting, and circulation statistics inquiries. Once the proper data base is established for the system, the computer is able to generate automatically the following types of reports and documents: circulation listing; book-overdue notice; book returned; fine-due notice; fine payment receipt; recall notice; hold notice; hold now-available notice; patron status report; reserve book list; payroll report; books usage statistics.

11. DELAYED-ENTRY SYSTEMS

A special class of data collection equipment is that which does not enter data directly to a computer or a central recorder but instead records it locally for subsequent transmission to a central location. This class can be grouped under the name "delayed-entry systems." It includes equipment that reads hand-encoded media as well as equipment for recording data on a local magnetic tape for later transcription to a central computer-compatible medium.

The advantages of delayed-entry systems are twofold: first, they permit off-line preparation and recording of source data. This enables source personnel who record the data to make sure it is correct before sending it and tying up communication or computing facilities and control recording as little as possible. Second, the equipment as a rule is versatile, portable, and inexpensive; sometimes the data entry instrument is a pencil! A practical means of subdividing delayed-entry equipment is to separate systems that read hand-encoded media from systems that read electronically encoded media.

HAND-ENCODED MEDIA SYSTEMS (OPTICAL READING EQUIPMENT)

Hand-encoded media delayed-entry data collection systems permit source personnel to record transaction data on paper documents. The three most important types of such equipment are mark-sense readers, optical character readers, and bar-code readers; all of them are also known by the common name "optical reading equipment." Each of the three types will be discussed by describing a representative commercial system.

Mark-Sense Readers: Motorola's MDR-Series Optical Readers

The Motorola MDR-series readers are desk-top devices that read key-punched, pencil- and typewriter-marked, or preprinted data in any combination from tab cards or other documents. Data is converted to one of several business machine codes and transmitted to a receiving device, such as a paper tape punch, card punch, magnetic-tape recorder, computer, or Teletypewriter set via such media as modems, Teletypewriter systems, or hardwire connections. Readers are available in models capable of operating in unattended, as well as attended, operational modes, enabling such operations as centralized polling of many remotely located readers.

Basic applications of the readers include inventory control, utility reading and billing systems, service reporting, and hospital record reporting. In each application, source data is typically hand-marked onto special forms preprinted in a special application-oriented format.

Another important manufacturer of mark-sense terminals is Hewlett-Packard.

Optical Character Readers: Cognitronics' ROCR System

Most optical character recognition (OCR) equipment is not practical for data collection purposes because OCR normally requires data to be printed in a special font, which requires equipment such as a typewriter; this requirement is impractical for data recording by source personnel. However, some OCR equipment can read hand-printed data—typically, numbers that are drawn according to special rules. Two types of hand-printed numeric schemes are shown in Figure 11-1. The numbers are entered on special forms in much the same way that marks are entered on mark-sense documents, except that instead of a coded representation of a number the actual number is drawn.

Most OCR equipment serves as a front end for a computer and is located in the same room; such equipment is not very practical for source data collection because manual document transportation is required. However, remote data entry terminals are being designed with hand-printed numerics recognition capability; such devices are offered by Cognitronics.

The Cognitronics remote optical character recognition (ROCR) system is the terminal for an OCR service bureau that utilizes a leased desk-top Model 720 remote scanner with the following capabilities:

1. Reads numeric pages, turnaround documents, journal tape, or continuous forms.

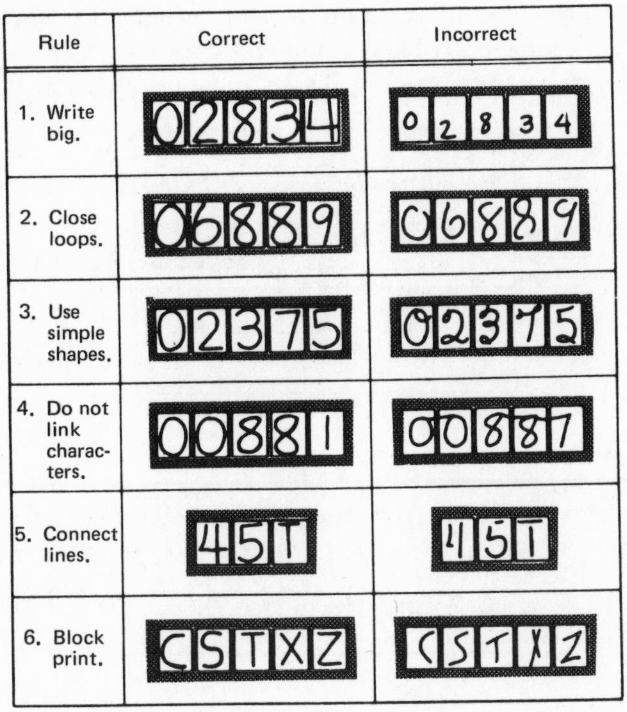

Rule	Correct	Incorrect
1. Write big.	02834	0 2 8 3 4
2. Close loops.	06889	06889
3. Use simple shapes.	02375	02375
4. Do not link characters.	00881	00887
5. Connect lines.	45T	45T
6. Block print.	CSTXZ	CSTXZ

Fig. 11-1.   Hand-printed numeric restrictions for IBM readers.

2. Recognizes the data at a central recognition facility.

3. Either records the data on seven- or nine-track magnetic tape to be delivered to the user or transmits the data to a remote seven-track recorder on the user's premises.

All common numeric fonts and a numeric hand-printed character set can be recognized; optional alphanumeric capability is also available.

Unreadable characters are displayed on a CRT at the Cognitronics recognition facility and are immediately identified by keyed entries. The system's headquarters uses a PDP-8 computer for control and Kennedy incremental recorders for recording. Cognitronics is marketing the service to small- and medium-sized users.

A special coded "Autoform" supplied by Cognitronics allows scanning to be controlled by preprinted coded instructions that designate fields to be read, fields to be accumulated, fixed- or variable-length fields, hand-printed or machine-printed fields, and check-digit fields.

Bar-Code Readers: Addressograph Series 9650 Scanners

Bar-code readers generally read documents that have been encoded by imprinting them with an embossed card. The card contains numeric data that is represented by a group of vertical lines (or "bars"), such as Addressograph's code, or by bars that are built into the numeric character, such as Farrington's code (see Fig. 11-2).

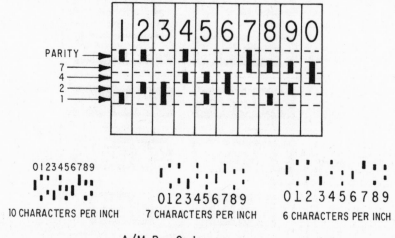

Fig. 11-2. Addressograph and Farrington numeric codes.

A typical application is charge account billing at gas stations, where sales slips are imprinted with the customers' credit cards by a simple $60 imprinter and then transported in a group to a processing center. The fact that manual transportation is required eliminates some of the advantages provided by true data collection, but these systems are so widely used that it is important to describe the processing equipment in some detail. True data collection equipment that can be used for, say, gas station applications will be discussed in Chapters 14 and 15.

The Addressograph Series 9650 scanners (Fig. 11-3) are optical code readers designed to process 80-column tab cards encoded with the Addressograph imprinted bar code, the Hollerith punched code, or combina-

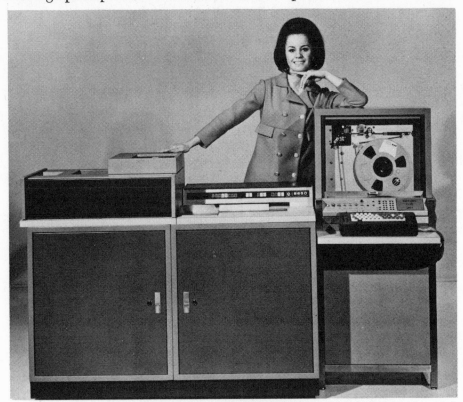

Fig. 11-3. The 9650 computer-oriented data entry scanner.

tions of both. The cards are converted to seven- or nine-track computer-compatible magnetic tape under programmed format control after undergoing parity and format checks. Checking and formatting occur while that data is stored in an intermediary buffer located in the tape unit, which is a Mohawk data recorder. A 64-character keyboard on the tape unit is used for error correction, and a printer on the scanner can be used for an audit trail. Data may enter the intermediary buffer directly from the 64-character data recorder keyboard or the tape drive as well as from the scanner, and may pass directly to the serial or line printer located on the scanner as well as to magnetic tape. Since the entire system may also connect to a data communications line, it can serve as a media printer, a keyboard-to-magnetic tape device, a communications terminal, or a computer output printer as well as a bar-code scanner.

The 9650 system used together with embossed card imprinters facilitate both data gathering at the source and preparation for rapid entry

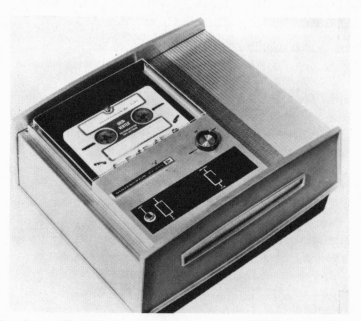

Fig. 11-4. Data-Verter system components (5023 magnetic-tape receiver not shown).

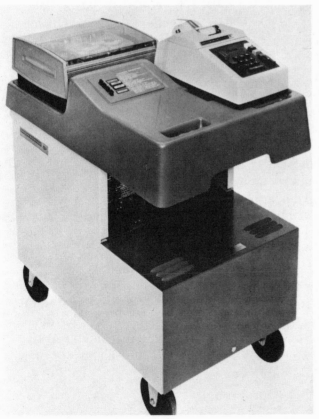

into a CPU via magnetic tape. Applications include data gathering in manufacturing for production and inventory control as well as for labor reporting, in hospitals to record and process patient treatment requests and test results, and in service industries for labor and service charge recording.

The input source for the card mode of operation is standard 80-column tab cards encoded with Addressograph numeric bar code, alphanumeric Hollerith punched code, or both. Bar-code density on the input documents can be six, seven, or ten to the inch as selected by the 9650 scanner program board. Up to three fields can be defined by the scanner program board for scanning bar-coded data and three fields for reading punched-hole data. Maximum reading speed is 300 cards per minute. Input hopper capacity is 1000 cards.

ELECTRONICALLY ENCODED MEDIA SYSTEMS

The second class of delayed-entry data collection systems is that in which individual devices record data electronically onto a self-contained medium such as a magnetic-tape cassette. The recorded data is later pooled with data from other similar devices and collected onto a common computer-readable medium such as ½-inch magnetic tape. One of the first complete systems of this type was the Digitronics Data-Verter remote system, which was introduced in 1966.

The Digitronics Data-Verter remote system is a data acquisition system employing a free-standing source, data magnetic-tape cartridge recording unit. After being recorded, the tape cartridge from the recording unit is transferred to an attended or unattended transmitter operating over voice-grade lines; the data is recorded on a nine-track magnetic tape or an eight-level punched tape terminal at the central computer facility. The noncomputer-compatible magnetic-tape cartridge is thus an intermediary medium between the source recording units and the computer-compatible medium, permitting the source recording units to be low in price and simple in design and operation.

The source-data recording units, which are either Model 708 or 709 adding machines, can be operated in cartridge record or nonrecord modes; all may be rendered completely mobile by the addition of the Model MC21 cart with its rechargeable battery. Main system components are illustrated in Fig. 11-4 and described in Table 11-1. The adding machine units are manufactured by Facit-Odhner, and all other system elements are produced by Digitronics.

Table 11-1.   Data-Verter System Components

COMPONENT MODEL	FUNCTION IN TOTAL SYSTEM	INDIVIDUAL COMPONENT I/O SPECIFICATIONS		
		INPUT	OUTPUT	COMMENTS
708 Adding machine with 900 recorder	Source data recording unit	10-key adding machine keyboard.	Printed listing tape; magnetic-tape cartridge (2-channel)	BCD or ASCII code; 10 figure listing, 11 figure totaling; 6 control keys. 708Z has leading zero feature
709 Adding machine with 900 recorder	Source data recording unit	10-key adding machine keyboard	Printed listing tape; magnetic-tape cartridge (2-channel)	Same capabilities as 708 with keyboard monitor indicator
800 Acoustic transmitter	Transmits data to central receiver via integral acoustic coupler	Magnetic-tape cartridge from 708 and 709 units	600 bps; public telephone lines	Attended transmission; 36 char/sec
802 Transmitter	Transmits data to central receiver via 202E data set	Magnetic-tape cartridge from 708 and 709 units	202E9 data set, 600 or 1200 bps; public telephones lines	Attended/ unattended transmission; 802-1 speed-36 char/sec; 802-2 speed-72 char/sec
803 Transmitter	Transmits data to central receiver via a data access arrangement	Magnetic-tape cartridge from 708 and 709 units	Data access arrangement; 1200 bps; public telephone lines	Attended, 803-1 unattended, 803-2
804 Acoustic transmitter	Transmits data to central receiver via integral acoustic coupler	Magnetic-tape cartridge from 708 and 709 units	600 or 1200 bps; public telephone lines	Attended transmission; 804-1 speed-36 char/sec; 804-2 speed-72 char/sec
5041 paper tape receiver	Records transmitted data on magnetic tape for computer input	202C2 data set	8-level punched paper tape	BCD or ASCII code; variable record length
5238 Magnetic-tape receiver	Records transmitted data on magnetic tape for computer input	202C2 data set	IBM-compatible, 9 track magnetic tape, 800 bpi	BCD numeric data written in BCDIC or ASCII code; 160 or 510 char block size

The Data-Verter system is designed as a relatively inexpensive but versatile source-data acquisition system. Although the line was originally developed for a large mail-order house, Digitronics is now concentrating

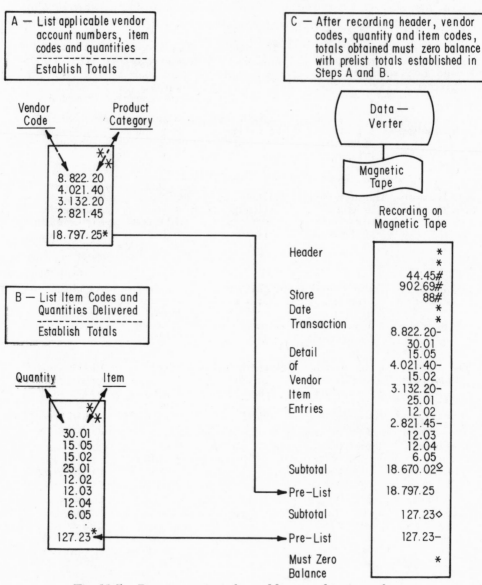

Fig. 11-5.   Representation of an adding machine tape from a typical ordering operation.

its marketing efforts on the food distribution market. A number of large supermarket chains are using Data-Verter systems to order items from their warehouses. Item names and their numbers and the level at which more should be ordered are printed on shelf labels. The battery-powered adding machine and recorder, mounted on a cart, are used to enter the item numbers and to order required quantities (a typical order format is shown in Fig. 11-5). Flexible code capability on the paper tape and magnetic-tape receivers provide for many other applications for the system.

Source data may be entered into the system through a Model 708 or 709 adding machine keyboard. The data keys are depressed one at a time, and then the ADD or SUBTRACT key is depressed. One digit is recorded on the magnetic-tape cassette as each data key is depressed. The ADD and SUBTRACT keys cause:

1. The line to be printed on the narrow listing tape.
2. A plus or minus symbol to be recorded after the entries on magnetic tape.
3. The entry to be added to or subtracted from the accumulator.

Ten digits plus a symbol can be printed on ADD/SUBTRACT entries. All possible operational keys, as well as the ten numeric keys record unique codes on the magnetic-tape cassette. The operational codes are commonly used as identifiers for types of transactions, location of transaction, end-of-transmission, and so forth. The CLEAR key, which causes previously entered data to be cleared from the buffer without printing, enters a special error code on tape to indicate the erroneous entry.

Eleven-digit totals are developed mechanically. Plus or minus totals are recorded on the printed adding machine tape but not on magnetic tape. The 708 and 709 contain individual total, subtotal, non-add, add, and subtract keys.

Tape cassettes recorded by the 708 and 709 may be used as input to the 800, 802, 803, and 804 transmitters. The two cassette models, C10 and C15, contain either 300 or 450 feet of 0.25-inch 2-track magnetic tape, respectively.

Data is recorded serially on one track at four characters per inch in a code configuration of five-bit ASCII or BCD for numeric data. The second track contains START/STOP and timing bits used for synchronizing the transmission of each character.

Data is received from the telephone line via a Bell Telephone 202C2 data set and is input to either a paper tape punch or a magnetic-tape receiver.

The 708 and 709 print all keyboard entries, subtotals, and totals on the adding machine listing tape. All subtract entries, negative subtotals, and negative totals are printed in red and all other entries in black. Note that when TOTAL or SUBTOTAL keys are operated, the total and subtotal is printed, but only the control code for that key is recorded on tape.

On the 709 adding machine, the keyboard monitor indicator lights momentarily each time data has been entered correctly and lights (and remains lit) if a key is not depressed fully or if two keys are depressed simultaneously. In addition, when an error condition is detected, the adding machine keyboard is locked.

Models 800 and 804 transmitters have an integral acoustic coupler that may be used with the public telephone network. Numeric data is transmitted at 600 bits per second (36 characters per second) using the 800 and 804-1 and at 1200 bits per second (72 characters per second) with the 804-2. Models 802 and 803 operate via a Bell System 202E data set and a data access arrangement, respectively. Data is transmitted at 600 bits per second using the 802-1 and at 1200 bits per second using the 802-2 and 803 transmitters. Model 803-1 requires attended transmission; Models 803-2 and the 802 operate in an unattended mode. The transmission code is usually ASCII, but numeric data may also be transmitted in a BCD code configuration. All transmitter models respond to the return call of the receiving station in initiating the data made, to prevent data from being sent to a "wrong number."

Data received from a 202C2 data set may be punched on eight-channel paper tape in preparation for computer entry; BCD numeric data can be punched in eight-level Flexowriter or ASCII code. Numeric data transmitted in ASCII is always punched in ASCII. The continuous data stream received from the communication line is recorded at 800 bits per inch in 160- or 510-character block length, and a single character TAPE MARK block is written at the end of a fully received transmission.

Many other cassette-recording/transmitting systems are available. Universal Data Acquisition offers data collection equipment that includes a 5½-pound, battery-operated, alphanumeric cassette recorder containing 16 keys and a strip printer. Other systems are offered by Electronic Laboratories and MSI.

12. COMMUNICATIONS-ORIENTED SYSTEMS

Data communications equipment provides means of performing data collection equivalent in function, but not in configuration, to previously described systems. This class of equipment generally does not provide integrated data collection systems, but proper equipment arrangements can provide for remote data entry and centralized data reception. The equipment comprises many types of data entry terminals that transmit, exclusively over data communications facilities (usually telephone lines), to data reception equipment comprising receiver/recorders, concentrators, and a communications processor.

Some data communication devices are much more practical for data collection functions than are others. Practically any reception device that can assemble from a number of remote locations is appropriate, but the most appropriate types of data entry terminals are those that provide for transaction data entry from source personnel. Thus the terminals must be designed to permit entry of data concerning a transaction and must be operable by personnel who have little or no training or skill in using such equipment. Two types of terminals best meet these conditions: those with an extremely simple design with a small number of keys and controls, and those designed for specific groups of applications, with some means of guiding the operator in accomplishing correct data entry for specific transactions.

PUSHBUTTON TELEPHONE DATA ENTRY

About the simplest type of data communications input device is the pushbutton telephone handset offered by the Bell System under the trade-name of Touch-Tone (Fig. 12-1). The advantages of using an ordinary

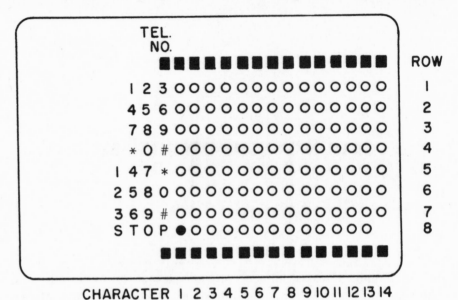

Fig. 12-1.   Bell System dial cards.

telephone handset are its wide availability, ease of use, and low cost. Its applications are unlimited. For example, sales personnel place orders to a central computer by calling the computer from any remote location equipped with a Touch-Tone telephone and then pushing the appropriate buttons representing the code numbers of the items to be ordered. As another example, retail store personnel update inventory and customer credit files by entering data through telephone handsets. In many cases, the receiving equipment provides audio responses to inquiries made from the handsets, in addition to receiving Touch-Tone signals converting them and formatting them for computer input. This function is widely used by retail stores for credit verification and by banks for customer account status inquiries. Receiving and response equipment will be described later in this chapter.

Pushbutton Telephone Construction

The Touch-Tone handset generates and transmits particular frequencies or tones in response to the depressing of its buttons. Each digit is represented by a combination of two characteristic tones. The length of each tone pulse is constant, regardless of how long a button is held down.

A standard handset provides 12 buttons, corresponding to the digits 0–9, plus two special function buttons. A 10-button model is also in service.

A limited amount of data prepunched into a dial card can be transmitted at a speed of approximately 10 characters per second. The dial card (Fig. 12-2) is perforated so that the punches can be easily pushed

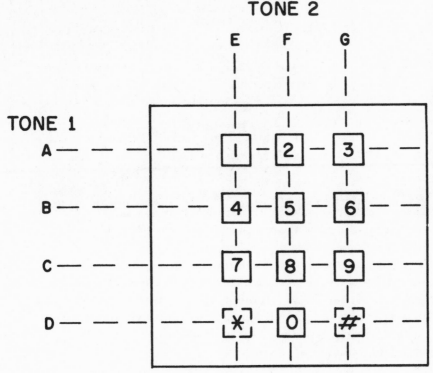

Fig. 12-2.   Touch-Tone pushbutton arrangement.

out by hand and read by the card dialer, which is normally used for semi-automatic dialing. After the dial card has been inserted into the dialer, it will rise until a stop punch is detected. Depressing the start key causes the dial card to be read for transmission. Multiple stop punches permit separate data fields in the dial card. Variable data, entered by means of the pushbuttons, is transmitted between each of the fixed data fields; blanks are not transmitted.

The dial card is used in many data collection applications to dial the receiving computer and then to enter semiconstant code data representing the transaction type and employee number as well as other such data. The pushbuttons are then used for entry of variable data.

Figure 12-3 illustrates the arrangement of a 12-button set. Depression of a button results in the generation and transmission of two distinct tones,

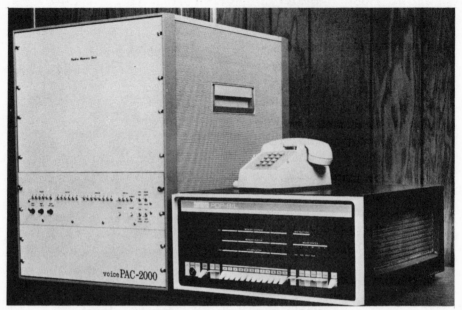

Fig. 12-3.   Periphonics voicePAC-2000 audio response system.

as indicated in the figure; for example, when the button marked 1 is pushed, frequencies A and E are transmitted. The tones A through G correspond to dial card rows 1 through 7. The two buttons marked * and # are present only on a 12-button handset or auxiliary dialer.

The following steps represent typical usage of the Touch-Tone handset in a data communications system:

1. The local agent or representative establishes a connection with the central office, using the telephone in the normal manner.

2. The local representative receives permission to transmit; the go-ahead signal may be verbal if the remote site is operating as an attended terminal or may be a distinct, audible tone if the remote site is operating as an unattended terminal.

3. The local representative transmits an identification code to allow the remote site to verify that a properly authorized person has placed the call; the remote site again transmits a go-ahead signal or disconnects, depending on the validity of the authorization code.

4. If authorized, the local representative transmits the data; this may consist of fixed data such as account number, prepunched into a dial card, plus variable data such as charge or order quantity.

5. The local representative transmits an end-of-message code.

6. Depending on the application and the facilities of the remote termi-

nal, the local representative may receive a verbal reply or audible signal indicating that the message was received in the proper format.

7. The transaction completed, the local representative hangs up; the remote terminal hangs up after a predetermined period of time elapses or in response to a special combination of codes transmitted before the local representative hangs up.

The range of applications is limited only by the imagination of the user. A few of the installed or planned applications include

1. Teller operations in a bank, including entering information into checking, mortgage, and loan accounts or checking the status of accounts.

2. Entering labor time and charge, material dispatching, production scheduling, or status reporting information in a manufacturing operation.

3. Central processing of sales orders entered directly from the salesman's home.

4. Central reply to inquiries entered from outlying business offices.

5. Central billing service for doctors.

PUSHBUTTON TELEPHONE DATA RECEPTION AND AUDIO RESPONSE

Touch-Tone telephones communicate over voice-grade telephone lines to stand-alone receiving systems or systems that serve as front ends for computers. Two such devices, which both also provide audio response functions, are described below.

Periphonics VoicePAC-2000 Audio Response System

The voicePAC-2000 (Fig. 12-4) is a buffered data communications unit that accepts multiple on-line inquiries, with standard pushbutton telephones normally functioning as the remote i/o devices. Under program control the voicePAC-2000 system relays on-line requests to the central processor; in the stand-alone configuration, input data is transferred to the system's own integrated data files or storage devices. In most data collection applications, the request or input data is used to update the appropriate data files. Once an inquiry is processed, an encoded message is returned to the voicePAC-2000, which interprets the message, assembles the appropriate sequence of vocabulary words, and transmits a voice response over the line to the terminal generating the original request.

Standard pushbutton telephones, rotary dial telephones equipped with auxiliary pushbutton pads, or special-purpose terminals serve as the remote i/o terminals in the voicePAC-2000 audio response system. Input

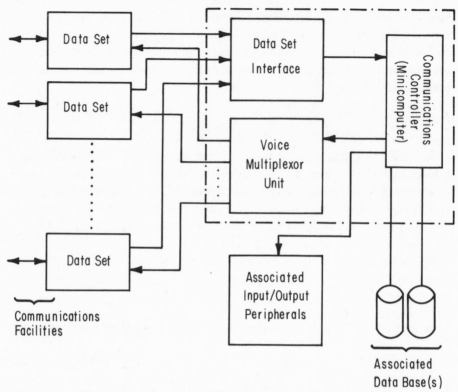

Fig. 12-4.   VoicePAC-2000 stand-alone configuration.

inquiries are transmitted using common-carrier line facilities. The incoming messages are received by Bell System Series 401 or 403 data sets and transferred to the data-set interface card within the voicePAC-2000. The input data is then multiplexed and routed to the voicePAC-2000 communications controller. The standard communications controller incorporates a Digital Equipment Corporation PDP-11 minicomputer. All organization and controlling functions, including lookup and buffering, are performed within the controller. Alternatively, other vendors' minicomputers may be incorporated within the controller to satisfy particular customer requirements.

The Periphonics voicePAC-2000 system contains an audio memory unit that assembles the prerecorded output message. The storage technique, which utilizes a high-speed magnetic disk, provides a wide range of vocabulary sizes. The standard unit can have a capacity of 32 to 2000 words, and optional configurations provide units that can generate in excess of 2000 words. The word length in the voicePAC system is completely unrestricted. Assuming a mean length of 0.5 second, a 128-word voicePAC-

2000 relates to 64 seconds (128 x 0.5) of contiuous audio output. Accordingly, a 2000-word voicePAC system provides in excess of 15 minutes of continuous audio output.

Through utilization of the advanced storage technique, rapid (less than 30 msec) random access to each word is provided for more natural sentence structuring of the appropriate response. The storage technique also allows access to any desired portion of a stored word. This fractionalization permits additional words to be derived from the existing stored vocabulary. For example, the word "information" actually affords access to four additional words: "in," "inform," "form," and "formation." If the word "automatic" is also in the vocabulary, it is possible under program control to combine syllable sequences and derive the word "automation" in conjunction with the preceding word example. This feature actually provides a vocabulary of greater capacity than that indicated in the model number of a particular voicePAC system.

In addition, the manner in which the audio response is stored on the magnetic disk facilitates the multiplexing of output lines. Identical or different messages can be directed simultaneously to all output lines accommodated by the system.

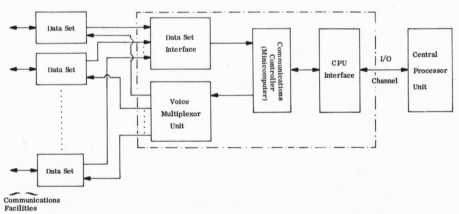

Fig. 12-5.   VoicePAC-2000 computer interface configuration.

The voicePAC-2000 audio response system can function either as a stand-alone unit or as a local or remote device interfaced to an existing central processing system. In the stand-alone configuration (Fig. 12-5), the core memory of the communications controller or an associated disk file unit(s) serves as the data base. Additional peripheral devices and input/output terminals may be interfaced to this arrangement to provide an expanded real-time system operating independently of the main processor.

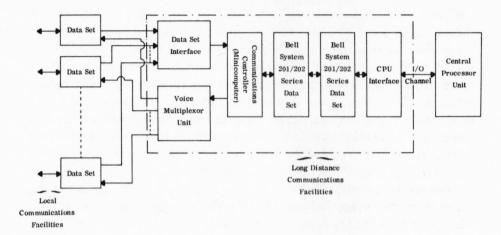

Fig. 12-6.   VoicePAC-2000 remote configuration.

In the computer-interfaced configuration, the interface to a central processing unit actually emulates a peripheral magnetic-tape control unit. This feature greatly minimizes any hardware or software considerations in interfacing to and communicating with the existing central processor. Additional on-line devices such as video displays and line printers can also be accommodated by this interface. In addition, programming may be

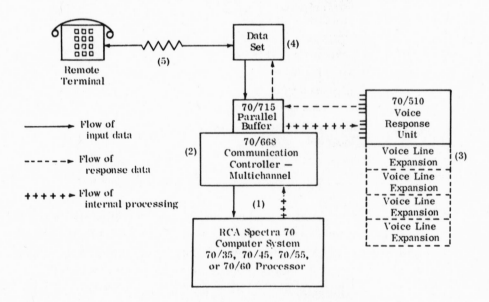

Fig. 12-7.   The RCA 70/510 voice response unit configuration possibilities.

performed in higher-order languages such as COBOL. Figure 12-6 depicts the interfaced arrangement.

In the remote configuration, the voicePAC-2000 (Fig. 12-7) can be connected to the central processor as a communications concentrator. Messages are concentrated at the remote location(s) in this mode. The interface requirements to the central processor are identical to those for the local configuration, with the exception that the appropriate modems must be substituted to accommodate the increased transmission rate. Data transmission rates up to 9600 bits per second can be accommodated by this system to reduce communications line requirements.

Due to its modular design, expansion of either the vocabulary or line capacity in the voicePAC-2000 system can be readily accomplished on site. For in-field modifications or additions to the stored vocabulary, words can be revised or inserted via totally automated techniques, using a standard tape recorder and an encoder module provided by Periphonics Corporation. Vocabulary no longer required in storage is eliminated under program control. The standard voicePAC-2000 system can accommodate up to 93 communications lines.

13. DATA COMMUNICATIONS TERMINALS
FOR DATA COLLECTION

Data communications terminals designed or capable of being used for data collection systems range from terminals very similar to a Touch-Tone phone to those as powerful and application-oriented as the specialized data collection input systems described in Chapter 9.

Most of these terminals are appropriate for the businessman who needs the benefits of a computer-oriented data collection system but who cannot afford his own system. Terminal-based data collection requires only an appropriate number of data entry terminals (often just one) and subscription to a central computer processing service center. Purchase prices are often in the range of $100 to $500 and rentals often are as low as $10 a month, an investment significantly less than the labor costs when using manual data entry methods.

The number of presently available terminals appropriate for data collection are far too many to allow detailed description of each, but they can be described adequately by summarizing the characteristics of representatives of the various types.

AUDAC DATA TERMINAL

The Audac Corp. data terminal (Fig. 13-1) is basically a Touch-Tone telephone equipped with a card reader, digital level switches, and an automatic prestored telephone number dialer. It is most often used to transmit credit sales data to a central computer with telecommunications

118

Fig. 13-1.   Audac data terminal.

facilities for credit authorization. In such applications, the operator enters the dollar sales amount on the lever switches, inserts a credit card encoded in a special Audac optical code, and presses a button that initiates automatic dialing to a central computer. When the computer responds, the operator presses another button to transmit the card and switch data. Additional variable data can be entered from the Touch-Tone buttons.

Computer actions depend on the specific system, but normally include accessing the appropriate customer credit file, checking the account status and trial current balance, and returning account status to the sending terminal via an audio response through the handset. If the checks pass, the computer updates the account. Although the Audac data terminal is primarily used for credit authorization, it can be used for many data collection applications such as production control, inventory control, and limited-access security systems. The card reader is capable of reading from 20 to 60 numeric digits from an Audac optically encoded data card.

COMPUTROL PORTABLE COMPUTER TERMINALS

Computrol Systems, Inc. offers several types of portable data entry terminals that can be configured for a variety of data collection functions. Most are battery-operated or a-c line-powered units built into an attaché case. The terminals contain an acoustic coupler into which an ordinary

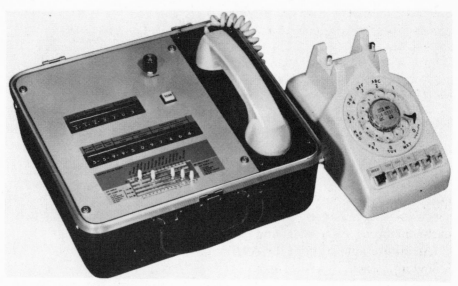

Fig. 13-2.   Computrol terminal for employment service data collection.

telephone handset is placed to communicate data to a computer via the
telephone system. The terminal operator sets data into the lever switches,
dials the computer with an ordinary telephone, places the handset into the
terminal's acoustic coupler, and presses a scan button. Data set into the
switches, plus data read by an optional card reader, is then sent to the
computer, which may send back data via audio response.

Terminals are specialized for specific application by switch arrange-

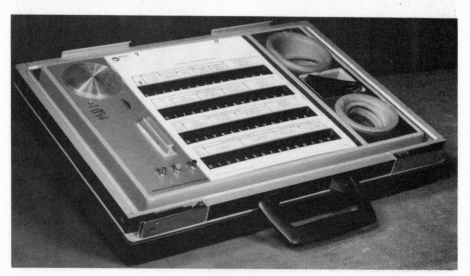

Fig. 13-3.   Computrol terminal for life insurance data collection.

ment and keyboard overlays. Figures 13-2 and 13-3 show Computrol terminals configured for collection of employment service and life insurance application data.

NATIONAL MIDCO MIDCODER 110 SYSTEM

The Midcoder 110 is an input data terminal primarily designed to perform credit verification and file updating for small businesses. The terminal comprises basically a card reader and variable dials whose contents are transferred automatically to a central computer via voice-grade telephone lines. Actual configuration of the terminal is variable and can be designed specifically for data collection operations from such establishments as department stores, restaurants and hotels, gas stations, and pharmacies.

As an example of how the Midcoder is used and how it works, the following account describes operating procedures for a Midcoder designed for use by gas stations: To complete a credit sale, the gas station attendant takes the customer's credit card to the Midcoder terminal, where he enters the total amount of the sale by adjusting digital lever switches, enters a transaction type code with other lever switches, and inserts the credit

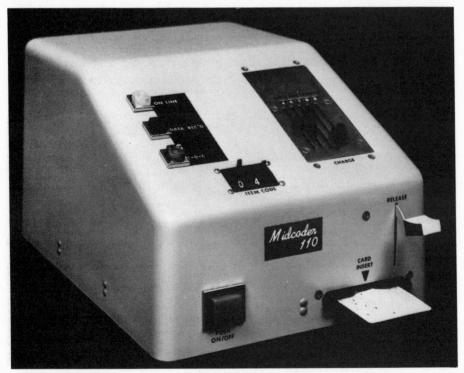

Fig. 13-4.   National Midco Midcoder 110 retailers terminal.

card. Upon card insertion, the terminal automatically dials a central computer to establish terminal-computer communication over conventional telephone lines. If the computer interface is busy, the terminal continues dialing until it is answered. Once communication is established, the terminal transmits a 5-digit sales amount, a 2-digit transaction code, the 12-digit account number encoded into the credit card, and a 9-digit location or terminal identification number prewired into the terminal.

The computer operates on the data in several different ways as it is received. The card number is first checked against an exception file that lists those accounts with bad credit, stolen cards, etc. If the check does not pass, the sending terminal is immediately notified by illumination of a lamp to signal such a condition. After checking the exception file, the computer checks the appropriate customer credit file to see if the current charge will cause the customer's credit total to exceed his credit limit. If charges do exceed the limit, the terminal is notified; if the limit is not exceeded, the computer updates the customer account file. This file can later be used directly for customer billing, since it is capable of automatically supplying account status data plus the date, amount, location, and type of each transaction in the billing period. The computer also updates separate files for the station and for the gas company. A separate file for taxes can also be maintained.

Operation and design of the Midcoder terminal are very similar to those for credit card sales from restaurants, hotels, and stores. Typical format for a transaction message from a retail store might include a 5-digit register/store location number, a 5-digit sales amount, an 11-digit credit card account number, a 7-digit item code, a 4-digit size, a 3-digit color code, and 3-digit tax amount. Separate files may be maintained for customer, store department, general inventory, store sales, and taxes.

National Midco also supplies a terminal for processing prescription data for pharmacies. The terminal accepts lever-switch entry of a drug code, prescription number, days' supply, charges, quantity, and customer identification. The computer checks for patient eligibility, excessive use, drug compatibility, and account status. The computer can maintain a number of files appropriate for the needs of the pharmacy and any creditor medical aid service it may subscribe to.

All Midcoder terminals may also be supplied with their own receipt printers, thus eliminating manual receipt preparation.

Such terminal-computer systems generate nearly all data required by businesses that use computer systems, drastically reducing paper work and extra labor that would normally be required. They are a step up for small businesses that already use a computer for transaction processing but which generate input data by manual or indirect methods.

CREDEX DATA TERMINALS

The Credex Corporation data terminals (Fig. 13-5) transmit lever-switch and credit-card-entered data to a computer and return computer responses via a numeric printer and a visual message display. The terminals

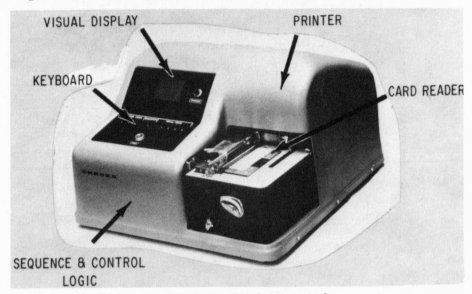

VISUAL DISPLAY   PRINTER

KEYBOARD

CARD READER

SEQUENCE & CONTROL LOGIC

Fig. 13.5   Credex data terminal.

are flexibly designed to meet the needs of particular applications. Libraries, retail stores, service stations, and hospitals use terminals equipped with, say, eight lever switches and a credit card reader, and factories use terminals equipped with a 15- to 20-lever switch keyboard, a 10-column badge reader, and an 80-column card reader. The terminals can be equipped to automatically dial a computer with teleprocessing facilities and operate in much the same way as the Midcoder terminals previously described. Responses from the computer are printed in a numeric printer or used to illuminate a message-display-panel, showing system status or the result of a computer action (see Fig. 13-6).

BURROUGHS TU 100 SERIES DATA COLLECTION AND INQUIRY TERMINAL

Late in 1970 the Burroughs Corporation released its TU 100 Series (Fig. 13-7) of terminals intended for low-cost on-line collection. The terminal provides for variable-data entry with a 10-key numeric keyboard and

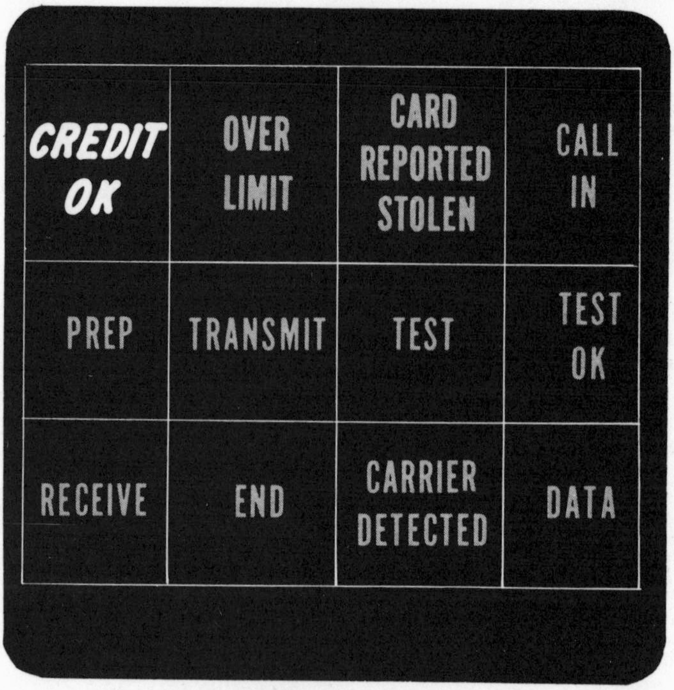

Fig. 13-6.   Credex message display panel.

fixed data entry with optional tab card and badge readers. Four function keys permit the operator to select the type of transaction to be entered, permitting the computer to select its proper operating program before the data is transmitted. Three other keys are used for communications control. Four terminal lights signal system and transaction status. The terminal also contains a printer, which can be used to print computer-generated alphanumeric output on a one- or two-part document or a continuous paper tape strip. Any one of eight printing line locations can be selected for document printing by using a keyboard control.

The TU 100 system can be configured in four basic ways:

1. *Direct Connection.* Up to nine terminals are connected directly to a computer communications line via a single two-wire pair.

2. *Switched-Line Connection.* A terminal is connected to a central computer via an ordinary switched telephone line with its own modem or acoustic coupler.

3. *Leased-Line Connection.* A "master" terminal is connected to a central computer via a leased telephone line with up to four "slave" terminals serially connected to the "master."

4. *Terminal-Unit Controller Connection.* Up to 20 terminals are connected to a Burroughs TU 920 terminal unit controller, which connects to a

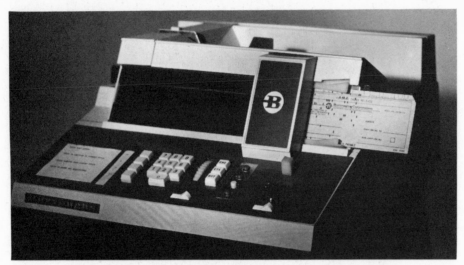

Fig. 13-7.   Burroughs TU-100 data collection terminal.

central computer via a single communication line and data sets. The TU 920 buffers and performs error checks on the incoming data.

In any of the configurations, the number of terminals is limited to the number of communications lines the computer is equipped to handle, multiplied by the number of terminals per line permitted by the configuration.

14. COMMUNICATIONS-RECEIVING AND -PROCESSING EQUIPMENT

There are two broad classes of equipment that can perform the receiving, recording, and processing aspects of data collection: receiving data terminals and communications processors.

DATA TERMINALS FOR RECEIVING PURPOSES

The complement of a transmitting-data entry terminal is a data-receiving terminal. These connect to communications lines and receive any compatible data transmitted over that line. Most perform limited control functions and record the incoming data onto punched paper tape or magnetic tape. Many receiving terminals are actually transmitter/receivers, capable of transmitting data from a medium such as punched paper tape or of recording incoming data into the same type of medium. Typically, they work in pairs, but can be used to receive data from source-data entry terminals as long as the communication format, codes, speed, and other characteristics are compatible.

COMMUNICATIONS PROCESSORS

Communications processing equipment include two basic system types, free-standing systems and multiline controllers.

Free-standing systems contain memory, processing capabilities, and input-output facilities. They are usually designed for a fairly specific purpose, such as message switching, and may have limited data processing capabilities. The systems may or may not be capable of exchanging data directly with a nearby general-purpose computer system.

Multiline communications controllers provide general-purpose computer systems with the capability of being connected to multiple communications lines. These controllers usually provide the control functions necessary to handle multiple simultaneous data paths, and the associated computer systems provide the memory, processing capability, and input-output facilities.

Primary characteristics of communications processors are the capability to control multiple data communications lines by, for example, automatically establishing a connection, both when transmitting and when receiving, and the capability to interpret data contained in incoming messages and assemble data for outgoing messages. A typical multiline controller and a free-standing system are described below.

Burroughs Multiline Control

A data communications subsystem for a Burroughs B2500, B3500, or B4500 computer consists of a peripheral control unit, the necessary adapters for control functions and line interfaces, and one or more remote terminal devices. Burroughs currently offers three different peripheral control units: the single-line control, the multiline control, and the terminal unit control. The multiline controls can communicate with various remote terminals at up to 9600 bits per second.

The B2500, B3500, and B4500 systems constitute Burroughs Corporation's current entry in the small- to medium-scale computer industry. The two systems are definitely "third generation" in their use of monolithic integrated circuits and read-only memory. Their processing power in business applications is impressive in terms of both internal speed and simultaneity. Moreover, Burroughs is placing a uniquely strong emphasis on multiprogrammed operation—and is supplying the hardware and software required to make multiprogramming (which Burroughs calls "multiprocessing") a practical reality.

There is no machine-language program compatibility between the B2500/3500 and the IBM/360, but the data formats, codes, and virtually all input/output media used in the B2500/3500 are IBM/360-compatible.

Multiline control (MLC) permits the connection of up to 36 narrow-band or voice-band communications lines to a single I/O channel. Only one MLC can be connected to a B2500, B3500, or B4500 computer system. A data communications subsystem built around a MLC includes: one MLC, one line adapter for each communications line, up to 36 communications lines (and data sets if required), and various remote terminal devices.

Collins C-System Free-Standing Communications Processor

The Collins C-system is a free standing, stored-program communications processor for terminating multiple telegraph-grade, voice-band, or broadband communications lines at standard speeds up to 50 kilobits per second.

The C-system consists of a family of hardware and software modules designed for continuous operation with fully automatic recovery from failures. Various functions include message switching, communications, and data processing. The software operating system employs the theories of queuing, load regulation and distribution, scheduling, input-directed processing, space allocation, batching, file organization, and file management. In parallel and in direct support of this software system, attention has been given to a hardware configuration that allows a large number of Teletypewriter terminals, man/machine I/O terminals, communications links of varying bandwidths, foreign processors, and user systems to be connected. The C-system is completely modular and expandable to accommodate the user who wishes to start with a smaller configuration, using only one computer.

This system provides hardware connection for up to 256 single-party or multiparty lines, in a half- or full-duplex mode, on any one computer. Full-duplex operation is recommended for lines in excess of 150 bits per second, to reduce turnaround times. The system structure allows for configuration of any number of computers required within a processing complex to terminate the desired number of peripheral devices and modem links.

The termination of peripheral devices (except for multiparty connections) provides for each device direct independent hardware access to core upon system allocation of core space to that termination. For multiparty connections, each device has direct independent hardware access to core upon system allocation of the communications facilities and core space. The MOS fabrication is used as an economical means of building the device termination units with sufficient logic to provide a balance between hardware functions during the movement of data into core and software functions following receipt of data in core.

Associated with each computer in the C-system is a variable number of zones of assigned secondary storage space. This storage utilizes one or more of the Collins disk files. The Collins disk file consists of 32×10^6 bytes of random access storage per drawer, has two independent drawers per standard cabinet, and has a 30-msec average access time, including latency and heat positioning time. All computers and secondary storage units reside on a common multichannel high-speed communications bus, which allows up to 64 independent and concurrent transactions between disk and computers. Communication of data between computers occurs via these secondary storage disk files. (One computer writes messages to its assigned disk zones, and the other computer reads the messages directly from the assigned area.) Supervisory messages (service messages), containing the storage address of data messages, are interchanged between processors as an indication that intraprocessor switching is required.

Important provisions in the design and implementation of the C-system are fault protection, fault detection, and failure recovery.

Features that provide failure protection include

1. Input regulation and control, with monitoring of available storage facilities and automatic control on all input message traffic to regulate the flow of data through the system within safe limits.

2. Redundant and symmetrical hardware structures (e.g., dual disk), allowing alternate processors to assume the load of failed units and alternate storage units to duplicate all significant data files, messages, and records; redundancy ensures no-break operation due to a single processor or disk failure.

3. Automatic transfer of all device connections from a failed processor to one or more designated alternates.

Features that provide failure detection include

1. Hardware logic monitoring of all processors, with automatic notification of failures to associated processors via the processor service units.

2. Routine responses from all peripheral devices and storage units indicating their status when interrogated.

Features that provide failure recovery include

1. Recovery programs to interpret the failure indicators received from failed units and initiate recovery action.

2. Generation of check-point data from pertinent data files and status records.

3. Automatic identification from any device that transfers to an alternate processor.

The net effect of the C-system input/output mechanism is a controlled environment for the input and output of traffic between processors and user terminals. This control allows input traffic to the system to be queued at the user terminal in peak traffic situations during which the system becomes short of arithmetic logic unit (ALU) time, core space, or secondary storage i/o capacity. This discipline adjusts the sizing of core, ALU time, and secondary storage throughout capacity to the average rather than the peak traffic loads. The user can reduce the probability of queuing at his terminal by minimizing the peak-to-average fluctuations in load to any one computer and by sizing the computer core and the number of computers in accordance with his expected traffic load. Withholding input service to a device allows positive control on system loading. For devices that supply "free wheeling" data to the system, a positive STOP code detector is required in the device adapter to allow the termination of input if insufficient core or ALU time occurs.

The multiplex (M) channel, in addition to performing the input/output functions described, provides switching service to all connected terminals. The C-system provides editing, routing, and queuing functions as a standard switching package for all terminal types. This package contains all features essential to communications and switching. Its routing function utilizes a "C-number" address, which is structured for optimum usage by a computerized network. The structure allows routing directories to be distributed throughout the network, with each processor accessing only that portion of the directory applicable to its local subscribers and the route to other areas or computers.

Other types of editing and routing are also executable within the M-channel as optional programs, and provide compatibility with some of the industry standards for people-to-people administrative message switching. The system is generally adaptable to any message format or link control procedure, although Collins recommends its own standards.

Integral parts of the standard switching and communications software package are the essential items of message logs, message retrieval, fault alarms, priority service, sequence numbering, line HOLD functions, intercept and reentry, possible duplicate and abort message labeling, reports, checkpointing, and dynamic routing change. The standard offering is augmented by a wide range of options that are consistent with many current user requirements.

15. TRENDS AND FUTURE DEVELOPMENTS

Many people see data collection increasing rapidly in importance in the next several years. The need for data collection arises both from the increasing dependence on computer processing and the increasing costs of creating source data with manual, indirect means. These additional costs are the result of the expense of inaccurate, slow, and incomplete data entry as well as the expense of increased labor.

Evidence of the increased importance and use of data collection is now indisputable. Large orders are being placed for data collection systems, especially in the field of retail merchandising. (For example, Sears is said to have placed a $100 million order in mid-1971 with Friden for its MDTS system.) Many other large orders are rumored to be pending between other retailers and manufacturers. Secondly, data communications is undeniably on the increase, data collection being an important aspect of data communications. All recent data collection equipment provides for either a data communications terminal or heavy use of data communications. The Carterfone decision (permitting non-Bell System equipment to interface with the public telephone network) and the construction of two major private communications networks (Data Transmission Company (Datran) and MCI Communications Corporation) provide the means and show the need for growth in data communications and thus in data collection systems.

APPENDIX: MANUFACTURERS DIRECTORY

Addressograph/Multigraph Corp., 1200 Babbitt Rd., Cleveland, Ohio 44117

American Regitel Corp., 589 Boston Post Rd., Madison, Conn. 06443

Audac Corp., 175 Bedford St., Burlington, Mass. 01803

Burroughs Corporation, Second Avenue at Burroughs, Detroit, Mich. 48232

Cognitronics Corp., 333 N. Bedford Rd., Mount Kisco, N. Y. 10549

Colorado Instruments, Inc., One Park St., Broomfield, Col. 80020

Computone Systems, Inc., Squires-Sanders Div. (*see* Computrol Systems, Inc.)

Computrol Systems, Inc. (formerly Computone Systems), 361 N. E. Paceo Ferry Rd., Atlanta, Ga. 30305

Control Data Corp., 8100 34th Ave. South, Minneapolis, Minn. 55420

Credex Corp., 7920 Charlotte Dr., S.W., Huntsville, Ala. 35802

Data Pathing, Inc., 370 San Aleso Ave., Sunnyvale, Calif. 94086

Dennison Manufacturing Co., 300 Howard St., Framingham, Mass. 01701

Digitronics Corp., 1 Albertson Ave., Albertson, N. Y. 11507

Electronic Laboratories, Inc., 3726 Dacoma St., Houston, Tex. 77018

Honeywell Data Products Division, 8611 Balboa Ave., San Diego, Calif. 92112

Information Machines Corp., 8811 Cuyamaca St., Santee, Calif. 92071

International Business Machines (IBM), Data Processing Machines, 112 East Post Rd., White Plains, N. Y. 10603

Kimball Systems, Inc., Division of Litton Industries, 151 Cortlandt St., Belleville, N. J. 07109

Mohawk Data Sciences, P.O. Box 630, Palisade St., Herkimer, N. Y. 13350

Motorola Instrumentation and Control, Inc., P.O. Box 5409, Phoenix, Ariz. 85010

MSI Data Corporation, 1381 Fischer Ave., P.O. Box 2193, Costa Mesa, Calif. 92627

National Cash Register Co., Dayton, Ohio 45409

Olivetti Underwood Corp., 1 Park Ave., New York, N. Y. 10016

Periphonics Corp., Box 244, Rte. 25A, Rocky Point, N. Y. 11778

Pitney-Bowes Inc., Walnut and Pacific Sts., Stamford, Conn. 06902

Ricca Data Systems, 1732 Reynolds, Santa Ana, Calif. 92705

Sierra Research Corp., 167 Bedford St., Burlington, Mass. 01803

Singer Company, Business Systems Division, 2350 Washington Ave., San Leandro, Cal. 94577

Uni-Tote, Division of General Instrument, 7400 York Rd., Baltimore, Md. 21204

Univac Data Processing Div., Sperry Rand Corp., P.O. Box 8100, Philadelphia, Pa. 19101

Univac Division, Sperry Rand Corp., P.O. Box 8100, Philadelphia, Pa. 19101

Universal Data Acquisition Co., Inc., P.O. Box 36166, 3928 Hartsdale Dr., Houston, Tex. 77036

GLOSSARY

Acoustic coupler. A modem designed for portable operation over the public telephone network whereby an acoustic and/or inductive connection is provided between the data terminal and the communications line, using a conventional telephone handset. Data to be transmitted is converted from a serial stream of binary digits to a sequence of tones (mark and space frequencies); at the receive end, each tone is converted to a stream of binary digits, corresponding to the original input data.

A/D converter. A device that transforms analog signals into digital signals.

Alphanumeric. Pertaining to a character set that includes both alphabetic characters (letters) and numeric characters (digits). *Note*: Most alphanumeric character sets also contain special characters.

ALU. Arithmetic logic unit.

Analog. Pertaining to data represented in the form of continuously variable physical quantities (e.g., voltage or physical position).

ASCII. American Standard Code for Information Exchange.

ASR. Automatic send/receive set. A combination Teletypewriter, transmitter, and receiver with transmission capability from either keyboard or paper tape; most often used in half-duplex circuits.

Audio response unit. Same as voice response unit.

Automatic check. A check performed by a facility that is built into equipment specifically for checking purposes. Also called built-in check, hardware check, or programmed check.

Baud. A unit of signaling speed equal to the number of discrete conditions or signal events per second. *Note*: In the case of a train of binary signals, and therefore in most data communications applications, one baud equals one bit per second.

BCD (Binary Coded Decimal). Pertaining to a method of representing each of the decimal digits 0 through 9 by a distinct group of binary digits. For example, in the "8-4-2-1" BCD notation used in many computers, the decimal number 39 is represented as 0011 1001 (whereas in pure binary notation it is represented as 100111).

Bit. A binary digit; a digit (0 or 1) in the representation of a number in binary notation.

Byte. A group of adjacent bits operated on as a unit and usually shorter than a word. *Note*: In a number of important current computer systems, the term "byte" has been assigned the more specific meaning of a group of adjacent bits, which represent one alphanumeric character or two decimal digits.

Call reception. The technique for answering calls originated by the remote data station; the method of reception can be manual, requiring operator attention, or automatic, whereby the receiving unit is usually capable of unattended operation.

Carrier. A signal suitable for modulation by an audio or other signal. The resultant signal can then be transmitted over a communications facility.

Cassette. A compact, permanently sealed, plastic container for a continuous computer i/o medium (such as magnetic tape or microfilm) that has been wound in a continuous loop.

Cathode-ray tube (CRT). A tube used to display data.

Central processor. The unit of a computer system that includes the circuits which control the interpretation and execution of instructions. Synonymous with cpu (central processing unit) and main frame.

Character. A member of a set of mutually distinct marks or signals used to represent data. Each member has one or more conventional representations on paper (e.g., a letter of the ordinary alphabet) and/or in data processing equipment (e.g., a particular configuration of 0 and 1 bits).

Character reader. A device that can translate human-readable characters directly into machine-readable characters.

Check digit. A digit associated with a word or part of a word for the purpose of checking for the absence of certain classes of errors.

Check-digit verification. A method of checking a critical number that has had an added check digit calculated partially by adding together multiples of all digits of the number in a fixed pattern.

CIRC. Circulation input recording center.

Code. A set of unambiguous rules that specifies the exact manner in which data is to be represented by the characters of a character set (e.g., ascii, Hollerith code).

Communications facilities. A path or group of parallel paths used for electrical transmission of data originating in digital form (also for voice and/or visual information) between two or more points. Some common types of communications facilities are telephone and telegraph cables, high-frequency radio, and line-of-sight microwave.

Communications front end equipment. Equipment located between a central computer and the communications line or lines to which it is connected. Performs the necessary conversion functions (speed, code, parallel-to-sequel) to allow efficient transmission of data between the computer and remote terminals. Makes the combination of the transmission (i.e., communications facility) and remote terminals appear to the computer like local peripheral equipment.

Communications link. The physical means of connecting one location to another for the purpose of transmitting information between them (e.g., a telegraph, telephone, radio, or microwave circuit).

Communications processing equipment. A device that performs one or more communications control and/or processing functions (such as speed and code conversion) while accepting data from multiple transmission lines. It can perform preprocessing functions and supervise efficient information transfer to a larger processing facility or operate as a stand-alone unit.

Compatibility. The characteristic that enables one device to accept and process data prepared by another device without prior code translation, data transcription, or other modification. Thus one computer system is data-compatible with another if it can read and process the punched cards, magnetic tape, etc., produced by the other computer system.

Concentrator. Usually a remote, buffered device designed to accept and queue data simultaneously from multiple, low-speed input lines for selective transfer, according to a predetermined sequence, over one or more high-speed lines to a data processing facility.

Configuration. (System configuration). A specific set of equipment units, which are interconnected and (in the case of a computer) programmed to operate as a system. Thus a computer configuration consists of one or more central processors, one or more storage devices, and one or more input/output devices.

Constant. A quantity whose value does not vary. *See also* Variable.

Controller communications. This permits party-line techniques to be employed. Multiple sending and receiving units may share the same communication line. With this type of installation, several hundred remote units may be on line. This system permits a wide array of on-line units to be employed, such as window posting machines, adding machines, and Teletype machines. The central controller provides simultaneous communications and processing. Up to 100 messages may be flowing in and out of memory simultaneously—and this may be time-shared with the processing of an independent program. The control center provides built-in accuracy. All transmission is checked for accuracy. Improper signals are detected immediately. When transmission errors exist, the controller demand that the data be retransmitted. All data entering the computer memory has been edited to ensure its accuracy and completeness. Each segment of data entering memory is automatically identified as to its source. Whether the system employs private or party-line facilities, all data is routed to its proper destination by the central controller.

CPU. Central processing unit.

Data channel. A path or group of parallel paths used for electrical transmission of data originating in digital form between two or more stations by wire or radio, or a combination of both. Also called link, circuit, line, path, or facility.

Data communications. The transmission of data from one point to another.

Data set. A device that serves as modulator and/or demodulator. Synonymous with *modem*. Provides the appropriate interface between a communications link and a data processing machine or system.

Data transcription. Conversion of data from one medium to another without alteration of its information content. *Note*: The conversion may be performed by manual key-stroke operation, by a computer system, or by a specialized converter, and may or may not involve changes in the format of the data.

Data transfer. The movement of data from a source to a destination (e.g., from one storage location or device to another).

DCOS. Data collection operating system.

Digit. A single numeric character used to represent an integer; e.g., in binary notation, the character 0 or 1; in decimal notation, one of the characters 0 through 9.

Direct distance dialing. An exchange service that enables a telephone user to specify subscribers outside the originator's local area.

EBCDIC (*Extended Binary Coded Decimal Interchange Code*). An 8-bit code that represents an extension of the 6-bit BCD code widely used in computers of the first and second generations.

EDP. Electronic data processing.

Embossed plate. A small metal or plastic card like a credit card with raised letters and numbers which can be used for imprinting.

FDM. Frequency-division multiplexer. *See* Multiplexers/concentrators.

File. A collection of related records, usually (but not necessarily) arranged in sequence according to a key contained in each record. *Note*: A record, in turn, is a collection of related items, whereas an item is an arbitrary quantity of data that is treated as a unit. Thus in payroll processing, an employee's pay rate forms an item, all items relating to one employee form a record, and the complete set of employee records forms a file.

Fixed input. A set series of codes (usually hardwired) automatically added to each data record.

Flip-flop: A device that has two states (e.g., a toggle switch).

Font. A family or assortment of graphic character representations (i.e., a character set) of a particular size and style (e.g., font E-13B, the MICR font adopted as a standard by the American Bankers' Association, and the U.S.A. standard optical font for OCR).

Form. (1)A printed or typed document, usually containing blank spaces for the insertion of specific data items. (2) Stationery on which data is printed for human use by automatic data processing equipment (e.g., by a line printer or tabulator).

Full duplex (Duplex). Pertains to the simultaneous, independent transmission of data in both directions over a communications link. *See also* Half-duplex; Simplex.

Graphic. A symbol produced by a process such as handwriting, drawing, or printing.

Half-duplex. Pertaining to the alternate, independent transmission of data in both directions—but in only one direction at a time—over a communications link. *See also* Full duplex; Simplex.

Hard copy. Pertaining to documents containing data printed by data processing equipment in a form suitable for permanent retention (e.g., printed reports, listings, and logs), as contrasted with volatile output such as data displayed on the screen of a cathode-ray tube.

Indicator. (1) A device that can be set into a prescribed state, often according to the results of a previous process, and which can subsequently be used by a control unit to determine a selection from alternative processes; e.g., an overflow indicator is set whenever an overflow occurs. (2) A device (e.g., a lamp) that informs an operator of the existence of a particular condition (e.g., power on, stacker full, hopper empty).

Input. (1) The process of transferring data from external storage or peripheral equipment to internal storage (e.g., from punched cards or magnetic tape to core storage. (2) Data that is transferred by an input process. (3) Pertaining to an input process (e.g., input channel, input medium). (4) To perform an input process. (5) A signal received by a device or component. *Note*: As the preceding definitions indicate, input is the general term applied to any technique, device, or medium used to enter data into data processing equipment, and also to the data so entered.

Input/output (i/o). A general term for the techniques, devices, and media used to communicate with data processing equipment and for the data involved in these communications. Depending on the context, the term may mean either input *and* output or input *or* output.

Inquiry. A method of interrogating the contents of computer storage from a keyboard.

Inquiry station. An input/output device that permits a human operator to interrogate a computer system and receive prompt replies in a convenient form. *Note*: Frequently, the inquiries are entered from a keyboard and the computer-generated replies are typed and/or displayed. Inquiry stations may be located remotely from the computer. An airline reservation system, for example, usually includes multiple inquiry stations in widely scattered locations.

Interface. A shared boundary (e.g., the boundary between two systems or between a computer and one of its peripheral devices).

I/O. Input/output.

Interlock. A protective facility that prevents one device or operation from interfering with another (e.g., by locking the keys of a console typewriter to prevent manual entry of data while the computer is transferring data to the typewriter).

Keyboard entry. Access to or entrance into an automatic data processing system by manual keys; an item of information so entered.

Line printer. A printer that prints all the characters comprising one line during each cycle of its action. *Note:* Two widely used types of line printers are chain printers and drum printers.

Management information system (MIS). A system designed to supply the managers of a business with the information they need so that they can be informed of the current status of the business, understand its implications, and make and implement the appropriate operating decisions.

Mark sensing. A technique for detecting pencil marks entered by hand in prescribed places on punched cards or other documents. The marked data may be converted into punched holes in the same cards, recorded on another medium, or transmitted directly to a computer.

MDTS. Modular data transaction system.

Message. An arbitrary amount of information (e.g., a group of characters or words) that is transmitted as a unit.

Message swtiching. A technique for controlling the traffic within a data communications network; involves the reception of messages from various sources at a switching center, the storage of each message until the proper outgoing communications link is available, and the ultimate retransmission of each message to its destination or destinations.

MICR (Magnetic Ink Character Recognition). The automatic reading by machine of graphic characters printed with magnetic ink.

MID. Management interrogation device.

MIS. Management information system.

MLC. Multiline control.

Modems (data sets). A modem (modulator/demodulator) or data set is a device that provides the appropriate interface between a data processing machine and a communications line. It converts data originating in digital form into analog signals suitable for transmission over telephone lines (and vice versa).

MOS/LSI (Metallic Oxide Semiconductor/Large-Scale Integration). A low-cost, microminiaturized circuit-manufacturing technique just beginning to be used widely in electronic devices such as terminals and calculators.

Multiline controller. One type of communications processing equipment designed to connect a general-purpose computer system to multiple communications lines. These devices usually maintain the control functions necessary to handle simultaneous data paths; the associated computer system provides the memory, processing capability, and input/output facilities.

Multiplex. To transmit two or more messages simultaneously over a single channel or other transmission facility. This is accomplished either by splitting the channel's frequency band into two or more narrower bands (frequency division multiplexing) or by interleaving the bits, characters, or words that make up the various messages (time-division multiplexing).

Multiplexers/concentrators. A multiplexer is a device that subdivides the capacity of a communication line so that it can be simultaneously shared by many users. Typically, many (10 to 30) low-speed (under 300 bits per second) devices input to a multiplexer whose output feeds one voice-grade line. The multiplexer only combines the data; it does not alter or process it.

Two types are available: time-division multiplexers (TDM) and frequency-division multiplexers (FDM).

Negative credit verification. A method of credit checking that examines a list of account numbers of those whose cards have been stolen, who have bad credit, or who require special attention.

OCR (optical character recognition). The automatic reading by machine of graphic characters through use of light-sensitive devices.

Odd-even check. See parity check.

OEM. Original equipment manufacturer.

Off-line. Pertaining to equipment or devices that are not in direct communication with the central processor of a computer system. *See also* On line. *Note*: On-line devices are usually under the direct control of the computer with which they are in communication.

Output. (1) The process of transferring data from internal storage to external storage or to peripheral equipment (e.g., from core storage to magnetic tape or a printer). (2) Data that is transferred by an output process. (3) Pertaining to an output process (e.g., output channel, output medium). (4) To perform an output process.

Parity bit. A bit (binary digit) that is appended to an array of bits to make the sum of all "1" bits in the total array either always even (even parity) or always odd (odd parity).

Parity check. A check that tests whether the number of "1" bits in an array is even (even parity check) or odd (odd parity) check. Frequently called odd-even check.

PEPPER. Photoelectric portable probe/reader.

Plugboard. A perforated board used to control the operation of some automatic data processing equipment. The holes in the board (called hubs or sockets) are manually interconnected by means of wires terminating in plugs (called patchcords), which are inserted in the proper sockets to effect the various operational jobs. Frequently called control panel.

POS. Point of sale.

Positive credit verification. A method of credit checking that examines and updates all customer credit balances in order to verify that a customer is in good standing.

Prepared input. Any prerecorded or previously input medium such as a punched card, credit card, or magnetic tape.

Progammed check. See Automatic check.

Read-after-write check. A check on the accuracy of an output operation in which the data recorded on the output medium is read back and compared with the data that was supposed to be recorded. *Note*: Read-after-write checking is one of the most positive types of error checking; it may be performed automatically, as in most of the current magnetic-tape units, or as a separately programmed operation, as in many disk storage units.

Remote batch terminal. A terminal that permits batch processing of the input/output of all data for a specific task (e.g., payroll) to occur from a remote location. It consists typically of medium-speed peripheral equipment such as a card reader, line printer, and magnetic-tape transport together with all necessary communication and transmission logic.

ROCR. Remote optical character recognition.

SDA (source-data automation). The capture, in machine-readable form, of data describing events or transactions at the time and place where each event or transaction occurs. *Note*: SDA can greatly improve the overall efficiency of data processing operations by reducing the need for manual data-transcription operations and by decreasing the incidence of errors.

Simplex. Pertaining to a communications link that is capable of transmitting data in only one direction. *See also* Full duplex; Half-duplex.

SKU. Stock-keeping unit.

Software. The collection of programs and routines associated with a computer (such as assemblers, compilers, utility routines, and operating systems), which facilitate the programming and operation of the computer. *See also* Hardware.

Source-data recording. Recording that is accomplished as a by-product of an original transaction. A magnetic-tape cassette, for instance, can be recorded with data relating to a sales transaction when a salesman depresses the keys to generate the customer's receipt.

Source document. A document from which data is extracted (e.g., a document containing typed or handwritten data to be keypunched).

SPICE. Sales point information computing equipment.

Symbol. A character, group of characters, or ideograph that serves as a representation of something else by reason of relationship, association, or convention.

System configuration. See Configuration.

TDM. Time-division multiplexer. *See* Multiplexers/concentrators.

Telecommunications. The transmission of signals over long distances, such as by radio or telegraph. *See also* Data communications.

Teletypewriter. Trade name used by AT&T to refer to telegraph-terminal equipment.

Time sharing. (1) The use of a given device by a number of other devices, programs, or human users, one at a time and in rapid succession. (2) A technique or system for furnishing computing services to multiple users simultaneously while providing rapid responses to each of the users. *Note:* Time-sharing computer systems usually employ multiprogramming and/or multiprocessing techniques, and they are often capable of serving users at remote locations via a data communications network.

TRACOM. Transaction communicator.

Transaction code. One or more characters that form part of a record and signify the type of transaction represented by the record (e.g., in inventory control, the types of transactions would include deliveries to stock, disbursements from stock, orders, etc.).

Transcribe. To convert data from one medium to another without altering its information content; i.e., to perform a data transcription operation.

Translate. (1) To transform statements from one language (the source language) to another (the object language) without significantly changing the meaning or information content, as in an assembler or compiler. (2) To convert data from one code to another.

Transmit. To send data from one device or location to another.

Unit Record. (1) A record similar in form and content to other records but is physically separate (e.g., a record on a punched card). (2) Pertaining to equipment or techniques for dealing with unit records as described in (1), especially to punched card equipment.

Validity check. (1) In hardware, a check that determines whether or not a particular character is a legitimate member of the permissible character set. (2) In programming, a check that determines whether or not the value of a particular data item falls within the permissible limits (e.g., a man cannot work 800 hours in a month, and no month can have a day 32).

Variable. A quantity that can assume any of a given set of values. *See also* Constant.

Variable input. Input not previously prepared, which alters with each transaction. Such input may be entered via a keyboard or via dial and switch settings.

Voice (audio) response unit. Equipment located at the computer site (as a special kind of front end equipment), which provides the computer system with the capability of generating verbal (prerecorded human voice) responses to data transmitted from remote Touch-Tone telephones or other types of audio response terminals.

INDEX

144